SpringerBriefs in Computer Science

SpringerBriefs present concise summaries of cutting-edge research and practical applications across a wide spectrum of fields. Featuring compact volumes of 50 to 125 pages, the series covers a range of content from professional to academic. Typical topics might include:

- A timely report of state-of-the art analytical techniques
- A bridge between new research results, as published in journal articles, and a contextual literature review
- A snapshot of a hot or emerging topic
- An in-depth case study or clinical example
- A presentation of core concepts that students must understand in order to make independent contributions

Briefs allow authors to present their ideas and readers to absorb them with minimal time investment. Briefs will be published as part of Springer's eBook collection, with millions of users worldwide. In addition, Briefs will be available for individual print and electronic purchase. Briefs are characterized by fast, global electronic dissemination, standard publishing contracts, easy-to-use manuscript preparation and formatting guidelines, and expedited production schedules. We aim for publication 8–12 weeks after acceptance. Both solicited and unsolicited manuscripts are considered for publication in this series.
**Indexing: This series is indexed in Scopus, Ei-Compendex, and zbMATH **

More information about this series at http://www.springer.com/series/10028

Oge Marques

Image Processing and Computer Vision in iOS

 Springer

Oge Marques
Department of Computer & Electrical
Engineering and Computer Science
Florida Atlantic University
Boca Raton, FL, USA

ISSN 2191-5768   ISSN 2191-5776   (electronic)
SpringerBriefs in Computer Science
ISBN 978-3-030-54030-2   ISBN 978-3-030-54032-6   (eBook)
https://doi.org/10.1007/978-3-030-54032-6

This Springer imprint is published by the registered company Springer Nature Switzerland AG.
The registered company address is: Gewerbestrasse 11, 6330 Cham, Switzerland

To Ingrid, with eternal gratitude for her unfailing support.

Preface

This book was motivated by the synergy between two main driving forces in today's technological landscape: (1) the maturity and popularity of image processing and computer vision (IPCV) techniques and algorithms; and (2) the unprecedented success of iOS-powered mobile devices, such as iPhone and iPad. The combination of these techniques and skills has been driving the emerging era of *mobile visual computing*.

This book presents a technical overview of some of the tools and technologies currently available for developing iOS applications with IPCV capabilities, including the latest advances on the use of machine learning and deep learning to build intelligent IPCV apps. Its main goal is to provide the reader a guided tour of what is currently available and a path to successfully tackle this rather complex, but highly rewarding, task.

This book is targeted at developers, researchers, engineers, and students who might need a roadmap to navigate the ever-changing maze of languages, libraries, frameworks, and APIs involved in developing IPCV applications for iOS.

The following are some of the highlights of each chapter in this book:

- Chapter 1 introduces the field of mobile visual computing and provides supporting facts and statistics that motivate the development of iOS apps in this space.
- Chapter 2 provides an overview of the basic steps in image processing and computer vision as well as the main aspects of the process of developing iOS applications for IPCV, with a particular emphasis on libraries and frameworks related to acquiring, storing, processing, and displaying images.
- Chapter 3 introduces the Core Image library and shows the examples of using Core Image for image filtering and face detection.
- Chapter 4 provides a quick conceptual review of Machine Learning and Deep Learning as well as an up-to-date coverage of how to develop intelligent IPCV apps using Core ML and Create ML.
- Chapter 5 presents Apple's Vision Framework and shows how it can be integrated with Core ML to develop intelligent IPCV apps.

- Finally, Chap. 6 provides an overview of the widely popular OpenCV library, with a special emphasis on the integration of OpenCV into iOS apps.

At the end of each chapter, I have included useful resources for the reader who wants to go further and start developing IPCV apps for iOS. These include books, tutorials, videos, GitHub repos, and selected examples.

Image Processing and Computer Vision remain extremely relevant fields, whose techniques can help solve many problems and whose growth has been propelled by many recent developments, particularly the popularization of deep-learning-based approaches for most IPCV tasks, which—thanks to Core Image, Core ML, Create ML, and Vision Framework—have become relatively easy to incorporate into iOS apps. I expect that this book will fulfill its goal of serving as a preliminary reference on these topics. Readers who want to deepen their understanding of specific topics will find more than a hundred references to additional sources of related information throughout the book.

I would like to express my gratitude to Christian Garbin for his comments and suggestions on an earlier version of the manuscript; Nataliia Neshenko, Humyra Chowdhury, and Steven Alexander for their assistance with diagrams and figures; and Roshni Merugu for her diligent fact-checking and bibliography research.

I would also like to thank Paul Drougas and his team at Springer for their patience, kindness, and support throughout this project.

Boca Raton, FL, USA   Oge Marques
July 2020

Contents

Chapter 1
Introduction

1.1 Motivation

This book was motivated by two main factors: (i) the maturity and popularity of image processing and computer vision techniques and algorithms; and (ii) the unprecedented success of iOS-powered mobile devices, such as iPhone and iPad.

On one side of the equation lie the mature fields of digital image processing and computer vision, whose techniques and algorithms have left the realm of exclusively scientific pursuits and associated demanding requirements (notably, the need for expensive, specialized hardware and software) to reach the lives of computer users and developers of virtually all levels. Digital cameras became ubiquitous over the past few years, to the point that we inevitably think of film-based cameras (and tape- or disk-based camcorders) as a thing of the past. The wide availability of image editing, retouching, and processing software helped popularize imaging terms and increase the use of image processing techniques, e.g., posterization, gamma correction, or Gaussian blur, to name but a few. The area of computer vision also experienced a transition through which the most successful algorithms for frequently used tasks have become commodities, available through libraries, several of which are open and free for developers and users.

On the other side of the equation lies the iPhone, originally released by Apple in 2007, which has revolutionized the smartphone industry with more than 2.2 billion units sold worldwide since then [6]. The tremendously successful iPhone was later joined by Apple's electronic tablet, the iPad, which—since its release in 2010—has sold more than 425 million units [5]. In their latest generations, both the iPhone and the iPad have a wide range of rich capabilities, including built-in camera sensors, multi-touch screen, and powerful processors, which have made it possible to conceive and develop applications that perform rich and complex image processing and computer vision tasks "on the go".

© The Author(s) 2020
O. Marques, *Image Processing and Computer Vision in iOS*, SpringerBriefs in Computer Science, https://doi.org/10.1007/978-3-030-54032-6_1

From a developer's standpoint, three factors combine to make the iOS a very attractive development platform for creating image-rich applications: (i) the devices' rich capabilities for image and video processing; (ii) Apple's programming architecture and its frameworks and libraries for image and multimedia and—more recently—machine learning; and (iii) third-party support for iOS-based development through open APIs and libraries, such as OpenCV. As a result, more than 37,000 photography-related apps are currently available in the App store [2, 11], with specialized subcategories such as: camera enhancement, image editing and processing, image sharing, and image printing, among others.

1.2   The Age of Mobile Visual Computing

We live in a world where images and videos are, indeed, everywhere [20]! Recent statistics indicate that more than 3.2 billion people have a smartphone with camera [10]. Thanks to technological developments during the past 25 years, there has been a significant increase in the production and consumption of visually rich contents, including high-quality images and high-definition videos. This growth has been accompanied by a shift towards performing many of these tasks, much of the time, using mobile devices such as iPhone and iPad.

There is, however, a big mismatch between the processes of *producing* visual resources and *organizing* them for further cataloguing and consumption. Production tasks (such as capturing, producing, editing, sharing, remixing, and distributing visual content) have become easier thanks to an extensive array of highly-capable devices and powerful apps. Organizational tasks (such as annotating, tagging, making sense of, searching for, and retrieving visual content), however, remain difficult and expensive.

Image and Video Capture

Taking pictures and videos has become an easy and inexpensive task. The increasing popularity of smartphones has made it possible to carry a photo camera and video recorder (app) at all times. The resolution and quality of images and videos produced with a smartphone camera are appropriate for everyday needs and even acceptable for venturing into amateur photography. The popularization of using smartphone cameras can be attested by the birth of a new term: *iPhoneography*, defined as "the art of shooting and processing photos with an Apple iPhone" [21].

Image and Video Editing

It has never been easier to enhance and fix a photo or perform basic video editing (such as trimming a video clip) and today's smartphones provide built-in features for such tasks. If the functionality provided by the smartphone's operating system and built-in apps is not enough, there are plenty of options to choose from: it is estimated that 2% [2] of the 1.85 million apps available at the Apple App store [11] belong to the *Photo and Video* category, which translates to more than 37,000 iOS apps to chose from!

Image and Video Publishing and Sharing

There are tens of websites (with companion iOS apps) for photo hosting and sharing to choose from, including: Flickr, Google Photos, SmugMug, 500px, Dropbox, and Amazon Prime Photos.

On the video arena, YouTube is—by far—the larger and most popular website for hosting user-generated videos, with more than 2 billion monthly users [18], an order of magnitude larger than one of its closest competitors, Vimeo [1]. Astonishingly enough, more than 500 *hours* of video content are uploaded to YouTube every *minute* [17]. The platform, which was created in 2005 (and acquired by Google for $1.65 billion USD in late 2006), has become one of the most visited websites in the world and a global phenomenon. Approximately 60% of mobile users in the U.S. access YouTube via a mobile device (smartphone or tablet) [15].

When it comes to social sharing of photos, Facebook, Instagram, WhatsApp, and Snapchat are the leading apps—at least 350 million photos are uploaded to Facebook every day [4], and Snapchat, Instagram, and WhatsApp users (combined) share at least 7.5 billion photos each day [8, 12, 16]. Other noteworthy players in the space of social media sharing of images and videos include TikTok, Periscope, and DailyMotion.

Image and Video Retrieval

Despite many technological advances in the field of Visual Information Retrieval (VIR) [20], the tasks of searching and retrieving relevant and useful images and videos remain rather challenging, whether we are searching for photos and videos from our personal smartphone or the web at large. During the past few years, however, several content-based image retrieval (CBIR) prototypes have graduated from research labs and become commercial apps, several of which in the form of Mobile Visual Search (MVS) [19] apps.

MVS solutions leverage the increasing availability and popularity of smartphones with cameras which have provided a much-needed use case for the query-by-example (QBE) paradigm, which consists of showing an image to the app and ask for information related to that image, such as: (i) can you find more images that look like (or are related to) this? (ii) where can I buy this product?—or simply—(iii) what, exactly, is this? There are, after all, numerous everyday scenarios in which the example image is right in front of the user's eyes and it's extremely easy and natural to snap a picture and use it to trigger a query. Examples of MVS apps include CamFind [3], Google Lens (embedded in the Google app for iOS) and a growing number of apps in which the ability to perform visual searches is embedded into an app, such as Vivino [14] (for wine labels), and the Amazon mobile app (for several product categories, such as books, toys, clothes, accessories, and even office supplies).

Developers have a broad range of APIs available for embedding visual search capabilities to their apps and websites. Examples of available APIs include: CloudSight [7] (the search engine behind the CamFind app [3] and TapTapSee [13], a mobile camera app designed specifically for blind and visually impaired users, which identifies objects and speaks the identification result out loud), and Mobile-Search [9] (from TinEye). These APIs allow the development of customer-centric solutions for which the intelligent processing of visual information is a requirement.

1.3   Mobile Visual Computing and iOS

The field of computer vision is advancing more rapidly than ever before. Many sophisticated algorithms for image processing and analysis have become available as commodities, thanks to the popularization of libraries such as Core Image (see Chap. 3) and OpenCV (see Chap. 6).

The growing use of deep learning algorithms to solve computer vision problems has energized the field and advanced the state of the art in several areas, from image classification to object detection and tracking, from image denoising to semantic image segmentation. It is now possible to use sophisticated pre-trained models for tasks such as age estimation, gender classification, and many others to build intelligent mobile apps with relative ease (see Chaps. 4 and 5).

In summary, iOS-based devices have become a rich platform for the development of mobile apps with rich visual computing capabilities in a wide variety of domains, from entertainment, to healthcare, to education. Thanks to Core Image, Core ML and the Vision Framework, the iOS ecosystem allows developers to create amazingly rich apps with intelligent visual processing capabilities, whose complexity is conveniently abstracted behind pre-trained models and powerful libraries.

References

1. 20 Vimeo Statistics and Facts (2020). https://bit.ly/2O5W8wJ. Accessed: 2020-07-06.
2. Apple: most popular app store categories in June 2020, by share of available apps — Statista. https://bit.ly/3e1tKGU. Accessed: 2020-07-06.
3. CamFind App - powered by CloudSight.ai API. Visual Search & Image Recognition API. http://camfindapp.com/. Accessed: 2020-07-06.
4. Facebook by the Numbers (2020): Stats, Demographics & Fun Facts. https://bit.ly/38vrYfW. Accessed: 2020-07-06.
5. How Many iPads Have Been Sold? A Breakdown by Quarter. https://bit.ly/38wV3ri. Accessed: 2020-07-06.
6. How Many iPhones Have Been Sold Worldwide? https://bit.ly/2AxNuUI. Accessed: 2020-07-06.
7. Image Recognition API & Visual Search Results — CloudSight AI. https://cloudsight.ai/. Accessed: 2020-07-06.
8. Instagram by the Numbers (2020): Stats, Demographics & Fun Facts. https://bit.ly/2ZBM18s. Accessed: 2020-07-06.
9. MobileEngine: Mobile Image Recognition and Augmented Reality. https://services.tineye.com/MobileEngine. Accessed: 2020-07-06.
10. Newzoo's Global Mobile Market Report: Insights into the World's 3.2 Billion Smartphone Users, the Devices They Use & the Mobile Games They Play. https://bit.ly/38tXGKD. Accessed: 2020-07-06.
11. Number of apps in leading app stores — Statista. https://bit.ly/38yApqF. Accessed: 2020-07-06.
12. Snapchat by the Numbers (2020): Stats, Demographics & Fun Facts. https://bit.ly/2Z41EGx. Accessed: 2020-07-06.
13. TapTapSee - Blind and Visually Impaired Assistive Technology. https://taptapseeapp.com/. Accessed: 2020-07-06.
14. The Vivino App. https://www.vivino.com/app. Accessed: 2020-07-06.
15. U.S. reach of leading video platforms 2018 — Statista. https://bit.ly/321Xb9x. Accessed: 2020-07-06.
16. WhatsApp blog: Connecting one billion users every day. https://bit.ly/3iy2djH. Accessed: 2020-07-06.
17. YouTube: hours of video uploaded every minute 2019 — Statista. https://bit.ly/2ZKwnHI. Accessed: 2020-07-06.
18. YouTube Revenue and Usage Statistics (2020). https://bit.ly/2BLbQuJs. Accessed: 2020-07-06.
19. B. Girod, V. Chandrasekhar, D. M. Chen, N.-M. Cheung, R. Grzeszczuk, Y. Reznik, G. Takacs, S. S. Tsai, and R. Vedantham. Mobile visual search. *IEEE Signal Processing Magazine*, 28(4):61–76, 2011.
20. O. Marques. Visual information retrieval: The state of the art. *IT Professional*, 18(4):7–9, July 2016.
21. S. Roberts. *The art of iphoneography : a guide to mobile creativity*. Pixiq, Asheville, 2012.

Chapter 2
Image Processing and Computer Vision iOS App Development: The Basics

2.1 Image Processing, Image Analysis, and Computer Vision

Digital image processing can be defined as the science of modifying digital images by means of carefully designed algorithms running on a digital computer—which, for the sake of this book, is an iPhone or iPad.

Traditionally, image processing operations used to be classified in three levels [41]:

- *Low-level*: primitive operations (e.g., noise reduction, contrast enhancement, etc.) where both the input and output are images.
- *Mid-level*: extraction of attributes (e.g., edges, contours, regions, etc.) from images.
- *High-level*: analysis and interpretation of the contents of a scene.

Since there is no universal terminology to delimit the boundaries between *image processing*, *image analysis*, and *computer vision*, for the sake of this chapter we will adopt the following convention:

- **Image processing**: operations where the input is an image,[1] and the output is a modified version of the image. Examples of techniques and algorithms in this category include: denoising (Fig. 2.1), sharpening, blurring, pseudocoloring, etc.
- **Image analysis**: operations where the input is an image and the output is a *labeled image*, where specific regions, edges, or contours from the input image have been outlined. Examples of techniques and algorithms in this category include: semantic segmentation (Fig. 2.2), corner detection, and edge extraction, among many others.

---

[1]Depending on the algorithm, the input could also be a series of (2D or 3D) images and/or one or more videos. This is valid for all three categories described here.

© The Author(s) 2020
O. Marques, *Image Processing and Computer Vision in iOS*, SpringerBriefs in Computer Science, https://doi.org/10.1007/978-3-030-54032-6_2

(a)   (b)

Fig. 2.1 Image processing example—denoising: (**a**) input image; (**b**) output image

Fig. 2.2 Image analysis example—semantic segmentation: relevant portions of the image have been automatically assigned a (color-coded) label, such as *pedestrian*, *car*, *road*, or *sky*

- **Computer vision**: operations where the input is an image which is used by an algorithm to perform tasks, such as object detection, object recognition, image classification (Fig. 2.3), object tracking (across multiple frames of a video sequence), as well as answer questions related to the semantic contents of the image/video, e.g., how many people appear in this photo? where was it taken? etc.

When the distinction among the three categories is irrelevant we might refer to them collectively as IPCV (image processing and computer vision) or group them under the umbrella of *intelligent processing of visual information*. The IPCV

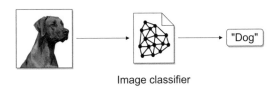

Image classifier

Fig. 2.3 Computer vision example—image classification: an image classifier (usually a pre-trained neural network) takes an image as input and produces the most likely label for that image

Fig. 2.4 IPCV pipeline example: original image (left), filtered (center), and subsequently cropped (right)

pipeline (or workflow) usually consists of several operations that are chained together in a meaningful sequence. For example, an image may be filtered and cropped before being used as an input to an object classifier (Fig. 2.4).

During the past two decades, thanks to the popularization of libraries, frameworks and toolboxes such as OpenCV [39] (for C++ and Python developers), MATLAB [24] (and its toolboxes[2]), ImageJ [20] and Fiji [16] (for Java developers), and many others, the building blocks of the IPCV pipeline have become commodities. More often than not—whether you are a developer, student, practitioner, or engineer—the fundamental operations behind each stage in your IPCV pipeline are immediately available through library functions. More recently, many of the most sophisticated algorithms for intelligent processing of visual information have also become available via cloud-based services and their APIs (e.g., Microsoft Azure Cognitive Services [8], Amazon AWS [1], Google Cloud AutoML [7]) which has led to enormous flexibility in building web-based vision applications.

Moreover, the impact of deep learning techniques on image processing and computer vision since the successful image classification results obtained by Krizhevsky et al. in ImageNet 2012 [40] has been so significant that there are deep-learning-based versions of virtually every IPCV task, often achieving state-of-the-art results for that task.[3]

---

[2]Notably the Image Processing, Deep Learning, and Computer Vision toolboxes.

[3]It is commonly said that the history of computer vision will be written in two volumes: (i) before deep learning (1950s–2012); and (ii) after deep learning (2012–present).

Most importantly for the sake of this book, the iOS ecosystem offers a rich platform for developing IPCV apps, thanks to several frameworks, libraries, and APIs (Sect. 2.3), particularly Core Image (Chap. 3), Core ML (Chap. 4), and the Vision Framework (Chap. 5). Moreover, the rich functionality of OpenCV can also be accessed by iOS apps (Chap. 6).

2.2   The iOS Development Environment

Xcode

Xcode [36] is the Integrated Development Environment (IDE) for designing, coding, debugging, and testing iOS apps. It includes a text editor where you write the source code that defines what your app does and a visual editor where you design your app's user interface [21].

Xcode also includes a compiler, a debugger, and a vast array of features for iOS app development. You can download Xcode for free from the Mac App Store. At the time of writing, the latest version of Xcode is Xcode 12, which assumes that you use Swift as a programming language and includes the new `SwiftUI` framework [21].

Swift

In the early days of iOS programming, the language of choice was Objective-C. Most of the code for iOS today is written in Swift. However, for legacy apps or IPCV apps that need to interface with OpenCV (see Chap. 6), knowledge of Objective-C and/or C++ is crucial.

The Swift programming language was announced in 2014 at Apple's World-wide Developers Conference (WWDC) and has become one of the most popular programming languages since then. It supports the object-oriented and functional programming paradigms that most programmers are comfortable with, and it also introduces a new one: protocol-oriented programming [21].

Swift prevents many common programming errors by adopting modern programming patterns, such as [32]:

- Variables are always initialized before use.
- Array indices are checked for out-of-bounds errors.
- Integers are checked for overflow.
- Optionals ensure that `nil` values are handled explicitly.
- Memory is managed automatically.
- Error handling allows controlled recovery from unexpected failures.

UIKit and SwiftUI

From the very beginning of iOS programming and up to iOS 12, iOS apps were written using the iPhone's original user interface framework, UIKit. UIKit was designed at a time when the general philosophy behind mobile app development was that mobile apps are like desktop apps, but on a less-powerful computer with a tiny screen. The overwhelming majority of iOS apps and their code are based on UIKit [21].

iOS 13 introduced SwiftUI, a new way for iOS developers to build user interfaces for their apps and to make it easier to port iOS apps to Apple's other platforms: macOS, watchOS and tvOS. If you're new to iOS development, experts recommend that you learn UIKit first and then transition to SwiftUI [21].

A Recipe for Success

To conclude this section, these are the suggested steps to learn the basic steps that should lead you to become a successful developer of IPCV iOS apps.

1. Get a Mac computer. You will need a Mac to run Xcode.
2. Download and install Xcode [36].
3. Sign up with Apple to become a registered iOS developer [2].
4. Learn the basics of iOS programming using Swift.[4]
5. Get an iPhone and/or iPad to test your apps.[5]
6. Learn about useful Apple frameworks, libraries, and APIs for IPCV tasks (see Sect. 2.3), particularly Core Image (Chap. 3), Core ML (Chap. 4), and the Vision Framework (Chap. 5).
7. Read as much as you can (books, blogs, Apple documentation, tutorials, etc.).
8. Practice, practice, practice.

2.3 Useful Frameworks, APIs, and Libraries

This section contains a list (in alphabetical order) of the most relevant Apple frameworks, APIs, and libraries for iOS developers working in IPCV and related areas

---

[4]This is a potentially time-consuming step, since it involves learning a new programming language *and* many software development aspects that are crucial to mobile app development. Fortunately, there are plenty of resources online to learn them, some of which appear at then end of this chapter.
[5]You can test much of the functionality of an iOS app using the iOS simulator that is bundled with Xcode, but for certain aspects of some apps (e.g., live video capture) you must deploy and test on a physical device.

(e.g., augmented reality, multimedia, computer graphics, games, video processing, and machine learning), with pointers to official documentation and additional information.

ARKit

One of the most exciting developments in mobile app development in recent years has been the growth of Augmented Reality (AR) [4]—often called *mixed reality*—apps, which allow the creation of "user experiences that add 2D or 3D elements to the live view from a device's camera in a way that makes those elements appear to inhabit the real world" [3].

ARKit is a framework that combines device motion tracking, camera scene capture, advanced scene processing, and display conveniences to simplify the task of building an AR experience, using the front or rear camera of an iOS device [3].

AVFoundation

The AVFoundation framework combines six major technology areas to support a broad range of tasks for handling audiovisual media on Apple platforms, namely [5]:

* *Media Assets and Metadata*: load, inspect, and export media assets and metadata, and perform low-level reading and writing of media sample data.
* *Media Playback and Selection*: get and inspect media assets; queue media for playback and customize playback behavior; edit and combine assets; import and export raw media streams.
* *Cameras and Media Capture*: capture photos and record video and audio; configure built-in cameras and microphones or external capture devices.
* *Media Composition and Editing*: combine, edit, and remix audio and video tracks from multiple sources in a single composition.
* *Audio Playback, Recording, and Processing*: play, record, and process audio; configure your app's system audio behavior.
* *Speech Synthesis*: convert text to spoken audio.

AVKit

AVKit is a framework that provides a high-level interface for playing video content, by allowing the creation of view-level services for media playback, complete with user controls, chapter navigation, and support for subtitles and closed captioning [6].

Core Graphics

The Core Graphics framework is based on the Quartz advanced drawing engine available for iOS, tvOS and macOS application development. Quartz 2D provides low-level, lightweight 2D rendering with unmatched output fidelity. Quartz 2D is resolution- and device-independent. You can use the Core Graphics framework to handle path-based drawing, transformations, color management, offscreen rendering, patterns, gradients and shadings, image data management, image creation, and image masking, as well as PDF document creation, display, and parsing [9].

The Quartz 2D API is easy to use and provides access to powerful features such as transparency layers, path-based drawing, offscreen rendering, advanced color management, anti-aliased rendering, and PDF document creation, display, and parsing [29].

Core Image

Core Image encapsulates image processing and analysis technologies that provide high-performance processing for still and video images. it includes many built-in image filters that can be used to process images and build complex effects by chaining filters [10]. Chapter 3 explores Core Image in greater detail.

Core Media

The Core Media framework defines the media pipeline used by `AVFoundation` and other high-level media frameworks found on Apple platforms. You can use Core Media's low-level data types and interfaces to efficiently process media samples and manage queues of media data [11].

Core ML

Core ML is a machine learning framework that allows developers to integrate machine learning models into their apps [12]. Chapter 4 explores Core ML in greater detail.

Core Video

Core Video provides a pipeline model for digital video. It simplifies working with video by partitioning the process into discrete steps, which makes it easier for developers to access and manipulate individual frames without having to worry about translating between data types or display synchronization issues [13].

Create ML

Create ML allows developers to create and train custom machine learning models to perform tasks such as recognizing images, extracting meaning from text, or finding relationships between numerical values [14]. Section 4.5 provides additional information on how to use Create ML to create intelligent iOS apps.

Image I/O

The Image I/O programming interface framework allows applications to read and write most image file formats. This framework offers high efficiency, color management, and access to image metadata [18].

ImageCaptureCore

The `ImageCaptureCore` framework, introduced in iOS 13, allows you to browse for media devices and control them programmatically from your app [19].

Metal

Metal is a framework for leveraging the computational power of graphics processors (GPUs) to quickly render graphics and perform data-parallel calculations. Many high level Apple frameworks are built on top of Metal to take advantage of its performance, including Core Image. Using one of these high-level frameworks shields you from the details of GPU programming, but writing custom Metal code enables you to achieve the highest level of performance [25].

Metal Performance Shaders

The Metal Performance Shaders framework contains a collection of highly optimized compute and graphics shaders that are designed to take advantage of the unique hardware characteristics of each GPU family to ensure optimal performance. It supports image processing operations, such as high-performance filters and histogram-based computations [17], as well as training neural networks for machine learning inference [33].

ML Compute

The ML Compute framework, introduced in iOS 14, can be used to accelerate training of neural networks across the CPU or one or more available GPUs. ML Compute uses the high performance BNNS primitives from the Accelerate framework for the CPU and Metal Performance Shaders for the GPU [26].

OpenGL ES

The OpenGL ES framework was one of the earliest 2D and 3D graphics frameworks in iOS. It was made available with iOS 2 and deprecated in iOS 12 in favor of the Metal framework [27].

PhotoKit

PhotoKit provides support for building photo-editing extensions for the Photos app, including direct access to the photo and video assets managed by the Photos app. Using PhotoKit, you can fetch and cache assets for display and playback, edit image and video content, or manage collections of assets such as albums, Moments, and Shared Albums [28].

VideoToolbox

The VideoToolbox is a low-level framework that provides direct access to hardware encoders and decoders, including services for video compression and decompression, and for conversion between raster image formats stored in CoreVideo pixel buffers [34].

Vision

The Vision Framework [37], introduced in iOS 11, encapsulates advanced capabilities for handling face detection and recognition, barcode and QR code detection, text detection, horizon detection, image registration (alignment), as well as object detection and tracking. It also interfaces with Core ML (Chap. 4) models for image analysis and machine learning tasks in the same workflow. Chapter 5 explores the Vision Framework in greater detail.

VisionKit

VisionKit is a small framework, introduced in iOS 13, that allow you to use the iOS camera as a document scanner [35].

2.4   Learn More About It

The official Apple site for Swift contains links to numerous books, sample code examples, videos, university courses, and official Apple documentation [31].

Ray Wanderlich's site is a rich source for high-quality books, tutorials, courses, and videos on iOS and mobile app development [30].

Since the early days of iOS programming, Stanford University's CS193p (Developing Applications for iOS) course has been a reference of high-quality materials and instructions. Slides, reading materials, and lecture videos are freely available online [15].

For a deeper look into design patterns and app architectures (and their limitations and alternatives), and how the same app can be implemented using a handful of different architectures, see [38].

Two special issues of the *objc.io* electronic magazine remain relevant for getting a deeper understanding of the development process behind mobile visual computing apps in iOS:

- Issue 21: Camera and Photos [22], which covers a wide range of topics, from image acquisition using the iPhone camera to selected frameworks, extensions, and GPU-accelerated apps.
- Issue 23: Video [23], which extends the discussion to video capturing, filtering, and hardware acceleration.

References

1. Amazon Web Services (AWS) Cloud Computing Services. https://aws.amazon.com/. Accessed: 2020-07-23.
2. Apple Developer Program — Apple Developer. https://apple.co/2WPfVoW. Accessed: 2020-07-23.
3. ARKit — Apple Developer Documentation. https://apple.co/2ZRcnEO. Accessed: 2020-07-14.
4. Augmented Reality — Apple. https://www.apple.com/augmented-reality/. Accessed: 2020-07-14.
5. AVFoundation — Apple Developer Documentation. https://apple.co/2ZVFyGN. Accessed: 2020-07-14.
6. AVKit — Apple Developer Documentation. https://apple.co/2WKmUQf. Accessed: 2020-07-14.
7. Cloud AutoML— Google Cloud. https://cloud.google.com/automl. Accessed: 2020-07-23.
8. Cognitive Services: APIs for AI Developers — Microsoft Azure. https://bit.ly/3eRsuGN. Accessed: 2020-07-23.
9. Core Graphics — Apple Developer Documentation. https://apple.co/2CK6ZdV. Accessed: 2020-07-14.
10. Core Image — Apple Developer Documentation. https://apple.co/30t8Xst. Accessed: 2020-07-14.
11. Core Media — Apple Developer Documentation. https://apple.co/2OWsOct. Accessed: 2020-07-14.
12. Core ML. https://apple.co/316Pf6h. Accessed: 2020-06-23.
13. Core Video — Apple Developer Documentation. https://apple.co/2EdBqto. Accessed: 2020-07-14.
14. Create ML. https://apple.co/2VekAQy. Accessed: 2020-06-23.
15. CS193p - Developing Apps for iOS. https://cs193p.sites.stanford.edu/. Accessed: 2020-07-28.
16. Fiji: ImageJ, with "Batteries Included". https://fiji.sc/. Accessed: 2020-07-23.
17. Image Filters — Apple Developer Documentation. https://apple.co/3eQiLjL. Accessed: 2020-07-22.
18. Image I/O — Apple Developer Documentation. https://apple.co/3jB7lE3. Accessed: 2020-07-23.
19. ImageCaptureCore — Apple Developer Documentation. https://apple.co/2WOgmjF. Accessed: 2020-07-23.
20. ImageJ. https://imagej.nih.gov/ij/. Accessed: 2020-07-23.
21. iOS Apprentice. https://bit.ly/30H3ZH5. Accessed: 2020-07-22.
22. Issue 21: Camera and Photos — objc.io. https://bit.ly/2EdCUUu. Accessed: 2020-07-14.
23. Issue 23: Video — objc.io. https://bit.ly/3jz7x6Z. Accessed: 2020-07-14.
24. MATLAB — MathWorks. https://www.mathworks.com/products/matlab.html. Accessed: 2020-07-23.
25. Metal — Apple Developer Documentation. https://apple.co/3hyMn7h. Accessed: 2020-07-22.
26. ML Compute — Apple Developer Documentation. https://apple.co/3fTEiJK. Accessed: 2020-07-22.
27. OpenGL ES — Apple Developer Documentation. https://apple.co/32N1KEX. Accessed: 2020-07-22.
28. PhotoKit — Apple Developer Documentation. https://apple.co/2BtnHxy. Accessed: 2020-07-22.
29. Quartz 2D: Introduction. https://apple.co/3eSsINS. Accessed: 2020-07-14.
30. raywenderlich.com. https://www.raywenderlich.com/. Accessed: 2020-07-28.
31. Swift — Resources — Apple Developer. https://apple.co/39kVMfN. Accessed: 2020-07-23.
32. The Swift Programming Language (Swift 5.3 beta) on Apple Books. https://apple.co/3fY8TWD. Accessed: 2020-07-23.

33. Training a Neural Network with Metal Performance Shaders — Apple Developer Documentation. https://apple.co/3jvB9Ca. Accessed: 2020-07-22.
34. VideoToolbox — Apple Developer Documentation. https://apple.co/2Ef7xcb. Accessed: 2020-07-22.
35. VisionKit — Apple Developer Documentation. https://apple.co/39pwEnT. Accessed: 2020-07-22.
36. Xcode — Apple Developer. https://apple.co/2CInyXt. Accessed: 2020-07-22.
37. Apple Vision Framework. https://apple.co/37V2bxg, 2018. Accessed: 2020-06-23.
38. C. Eidhof. *App architecture*. objc, Berlin, Germany, 2018.
39. A. Kaehler and G. Bradski. *Learning OpenCV 3: Computer Vision in C++ with the OpenCV Library*. O'Reilly Media, Inc., 2017.
40. A. Krizhevsky, I. Sutskever, and G. E. Hinton. Imagenet classification with deep convolutional neural networks. In *Advances in neural information processing systems*, pages 1097–1105, 2012.
41. O. Marques. *Practical Image and Video Processing Using MATLAB*. Wiley - IEEE. Wiley, 2011.

Chapter 3
Core Image

3.1 Introduction

Core Image is an image processing and analysis framework for iOS and macOS. Core Image provides a broad array of useful image processing and computer vision (such as face- and facial features detection in images and videos, and face tracking in videos) as well as almost 200 built-in image filters, organized into more than a dozen categories (Sect. 3.3) [7]. Moreover, thanks to its plug-in architecture, Core Image allows users to extend its functionality by writing custom filters that integrate with the system-provided filters [16].

3.2 Fundamental Classes

Core Image has three classes that support image processing on iOS [8]:

- `CIFilter` is a mutable object that represents an effect. A filter object has at least one input parameter and produces an output image.
- `CIImage` is an immutable object that represents an image, which can be synthesized from image data, read from a file, or produced at the output of another `CIFilter` object.
- `CIContext` is an object through which Core Image draws the results produced by a filter. A Core Image context can be based on the CPU or the GPU.

Listing 3.1 shows the fundamental steps for processing an image using Core Image methods specific to iOS. In line 2, a `CIContext` object is created with

© The Author(s) 2020
O. Marques, *Image Processing and Computer Vision in iOS*, SpringerBriefs in
Computer Science, https://doi.org/10.1007/978-3-030-54032-6_3

```
 import CoreImage
 let context = CIContext()
 let filter = CIFilter(name: "CISepiaTone")!
 filter.setValue(0.8, forKey: kCIInputIntensityKey)
 let image = CIImage(contentsOfURL: myURL)
 filter.setValue(image, forKey: kCIInputImageKey)
 let result = filter.outputImage!
 let cgImage = context.createCGImage(result, from: result.extent)
```
Listing 3.1 The basics of applying a filter to an image on iOS [8]

default options. The code in lines 3–4 creates a filter[1]and sets values for its parameters. In lines 5–6, a `CIImage` object representing the image to be processed is created (in this particular case, using a URL as its source). In line 7, the output image (another `CIImage` object) is produced (but not yet rendered). Finally, in line 8, the resulting image is rendered to a Core Graphics image, which can be displayed or saved to a file [8].

3.3 Filters

Within the context of Core Image, a *filter* is an object that has a number of inputs and outputs and performs some kind of transformation. For example, a sharpening filter might take an input image and an additional (scalar) parameter indicating the desired amount of sharpening and produce an output image with a crisper look. A *filter graph* is a network (directed acyclic graph) of filters, chained together so that the output of one filter can be used as the input of another; by chaining filters in this way, elaborate effects can be achieved [16].

Core Image comes with dozens of built-in filters, organized into categories.[2] A *filter category* specifies the type of effect—for example blur, color adjustment, or halftone—or its intended use—still images, video, non-square pixels, and so on. A filter can be a member of more than one category. A filter also has a *display name*, which is the name used to show to users and a *filter name*, which is the name used to access the filter programmatically. Since the list of built-in filters can change over time, Core Image provides methods that allow for querying the system for the available filters [8].

Most filters have one or more input *parameters*; each input parameter has an *attribute class* that specifies its data type, such as `NSNumber`. An input parameter can optionally have other attributes, such as its default value, the allowable

---

[1]In this example, the filter is `CISepiaTone`, which maps the colors of an image to various shades of brown.

[2]The official *Core Image Filter Reference* [5] lists the filters currently available, their characteristics and parameters, and shows a sample image produced by each filter.

```
CIFilter *myFilter =
    [CIFilter filterWithName:@"CIColorControls"];
NSDictionary *myFilterAttributes = [myFilter attributes];
```
Listing 3.2 Retrieving the attributes for a filter [8]

minimum and maximum values, the display name for the parameter, and other attributes described in *CIFilter Class Reference* [3]. Most filters have default values for each non-image input parameter [8]. Filter parameters are stored as key-value pairs. The key is a constant that identifies the parameter and the value is the setting associated with the key [8].

Querying the System for Filters (and Their Attributes)

To get a list of system filters, we can query Core Image for the names of the available filters in the kCICategoryBuiltIn category.[3] To know which input and output parameters are offered by a filter, we can ask for its inputKeys and outputKeys arrays of NSStrings, respectively. After a list of filter names has been obtained, we can retrieve the attributes for a filter by creating a CIFilter object and calling the method attributes which allows access to the attributes dictionary provided by the filter (Listing 3.2). Each input and output parameter name maps to a dictionary of its own, describing what kind of parameter it is, and its minimum and maximum values, if applicable [13, 16].

For example, here is the dictionary corresponding to the inputBrightness parameter of the CIColorControls filter [16]:

```
inputBrightness = {
    CIAttributeClass = NSNumber;
    CIAttributeDefault = 0;
    CIAttributeIdentity = 0;
    CIAttributeMin = -1;
    CIAttributeSliderMax = 1;
    CIAttributeSliderMin = -1;
    CIAttributeType = CIAttributeTypeScalar;
};
```

---

[3] The functionality for discovering which filters are available is not limited to built-in filters; it can be used for any category, such as kCICategoryVideo (which means the filter can be used for video) or kCICategoryBlur (which includes all blurring filters).

```
 1  func applyFilterChain(to image: CIImage) -> CIImage {
 2      let colorFilter = CIFilter(name: "CIPhotoEffectInstant",
        withInputParameters:
 3          [kCIInputImageKey: image])!
 4
 5      let bloomImage = colorFilter.outputImage!.applyingFilter("
        CIBloom",
 6
        withInputParameters: [
 7
        kCIInputRadiusKey: 10.0,
 8
        kCIInputIntensityKey: 1.0
 9          ])
10
11      let cropRect = CGRect(x: 350, y: 350, width: 150, height:
        150)
12      let croppedImage = bloomImage.cropping(to: cropRect)
13      return croppedImage
14  }
```
Listing 3.3 Example of a filter chain [8]

Filtering an Image: The Workflow

The process of filtering an image consists of three steps: creating and configuring a filter graph (which contains one or more filters to be applied sequentially to the input image), sending an image in to be filtered, and retrieving the filtered image [16]. Constructing a filter graph consists of instantiating individual filters, setting their parameters, and chaining them in the proper sequence to achieve the desired overall effect. Core Image optimizes the application of filter chains to render results quickly and efficiently, by making each CIImage object in the chain not into a fully rendered image, but instead merely a "recipe" for rendering. As a result, instead of executing each filter individually, wasting time and memory rendering intermediate pixel buffers that will never be seen, Core Image combines filters into a single operation[4][8].

Listing 3.3 shows an example of an image filter chain using Core Image, which applies a color effect to an image (CIPhotoEffectInstant), then adds an edge softening and glow effect to the result (CIBloom), and finally crops a section out of the result[5][8].

---

[4]In fact, Core Image can even reorganize filters when applying them in a different order will produce the same result more efficiently [8]

[5]See [12] for another example, if interested.

Using Special Filter Types

The majority of the built-in Core Image filters operate on a main input image and produce a single output image. There are several exceptions, though, which are listed below [8].

- Compositing (or blending) filters[6]combine two images according to a preset formula, e.g., the `CIMultiplyBlendMode` filter multiplies pixel colors from both images, producing a darkened output image.
- Generator filters[7]allow the creation of a new image from scratch, e.g., filters such as `CIQRCodeGenerator` and `CICode128BarcodeGenerator` generate QR code/barcode images that encode specified input data.
- Reduction filters operate on an input image, and produce descriptive information about the image, e.g., the `CIAreaHistogram` filter outputs information about the numbers of pixels for each intensity value (i.e., the histogram of pixel value counts) in a specified area of an image.
- Transition filters[8]can be used to create an animation that starts with one image, ends on another, and progresses from one to the other using an interesting visual effect, e.g., using the `CIDissolveTransition` filter to produce a simple cross-dissolve, fading from one image to another.

Creating Custom Filters

There are three ways by which the Core Image Filter class can be extended to create custom effects:

1. By chaining together several existing filters and creating new filters as a result of this compositional technique (see Chapter 5 of [15]).
2. By subclassing existing Core Image filters (see [14] for several recipes).
3. By writing a routine, called a *kernel*, that specifies the calculations to perform on each source image pixel (see Chap. 6 of [15]).

---

[6]For a complete list of compositing filters, you can query the `CICategoryCompositeOperation` category.

[7]For a list of generator filters, you can query the `CICategoryGenerator` and `CICategoryGradient` categories.

[8]For the complete list of transition filters, you can query the `CICategoryTransition` category.

3.4 Face Detection Using Core Image

Core Image offers a range of `CIDetector` objects that can analyze still and moving images and search for faces, rectangles, areas of text and Quick Response (QR) codes. The resulting detection is fast enough to work in real time with live video [15]. Detected features are represented by `CIFeature` objects that provide more information about each feature [2].

One of the most useful applications of `CIDetector` is face detection.[9] i.e. "the identification of rectangles that contain human face features" [11]. When Core Image detects a face, it also provides information about face features, such as eye and mouth positions, and can use the information about the face location to track its position in a video sequence.

Face detection can be used as a preprocessing stage in some compositional image processing tasks, such as [11]: (1) cropping or adjusting the image quality of the face (tone balance, red-eye correction, etc.); (2) selectively applying a filter (e.g., pixellate with `CIPixellate`) only to the faces in an image; and (3) placing a vignette around a face.

Listing 3.4 shows example code for detecting faces using the `CIDetector` class. Line 1 creates a context with default options. Line 2 creates an options dictionary to specify accuracy for the detector (in this case, high accuracy, which also implies slower processing). Lines 3–5 creates a detector for faces. Line 6 sets up an options dictionary for finding faces and sets the image orientation accordingly, so the detector knows where it can find upright faces. Finally, Line 7 uses the detector to find features in an image [11].

The input image must be a `CIImage` object. At the output, Core Image returns an array of `CIFeature` objects, each of which represents a face in the image.

```
1 CIContext *context = [CIContext context];
2 NSDictionary *opts = @{ CIDetectorAccuracy :
    CIDetectorAccuracyHigh };
3 CIDetector *detector = [CIDetector detectorOfType:
    CIDetectorTypeFace
4                                 context:context
5                                 options:opts];
6 opts = @{ CIDetectorImageOrientation :
        [[myImage properties] valueForKey:
    kCGImagePropertyOrientation] };
7 NSArray *features = [detector featuresInImage:myImage options:
    opts];
```
Listing 3.4 Face detection in Core Image [11]

---

[9]Face *detection* should not be confused with face *recognition*. Core Image does not contain any built-in method for face recognition.

```
for (CIFaceFeature *f in features) {
  NSLog(@"%@", NSStringFromRect(f.bounds));

  if (f.hasLeftEyePosition) {
      NSLog(@"Left eye %g %g", f.leftEyePosition.x, f.
      leftEyePosition.y);
  }
  if (f.hasRightEyePosition) {
      NSLog(@"Right eye %g %g", f.rightEyePosition.x, f.
      rightEyePosition.y);
  }
  if (f.hasMouthPosition) {
      NSLog(@"Mouth %g %g", f.mouthPosition.x, f.mouthPosition.y)
      ;
  }
}
```
Listing 3.5 Examining face feature bounds [11]

Listing 3.5 shows how to loop through that array to examine the bounds of each face and their characteristics, such as where the eyes and mouth are located [11].

3.5 Auto Enhancement Filters

Core Image includes filters for automatically enhancing images and correcting common issues found in photos, such as poor contrast or red eye due to camera flash. The auto enhancement feature of Core Image analyzes an image's histogram, face region contents, and metadata properties and returns an array of `CIFilter` objects whose input parameters are already set to values that will improve the analyzed image [1].

These are the filters Core Image uses for automatically enhancing images (and their purpose) [1]:

- `CIRedEyeCorrection`: Repairs red/amber/white eye due to camera flash
- `CIFaceBalance`: Adjusts the color of a face to give pleasing skin tones
- `CIVibrance`: Increases the saturation of an image without distorting the skin tones
- `CIToneCurve`: Adjusts image contrast
- `CIHighlightShadowAdjust`: Adjusts shadow details

The auto enhancement API has a method, `autoAdjustmentFiltersWith Options:`, which returns an array of options filters that can be chained together and applied to the analyzed image, as shown in Listing 3.6. In line 1 of the code, an options dictionary is created and the orientation of the image is used to set a value for the key `CIDetectorImageOrientation`. Line 2 illustrates how the `NSArray` of options is created. Finally, Lines 3–6 show how each filter is applied to the image [1].

```
NSDictionary *options = @{ CIDetectorImageOrientation :
    [[image properties] valueForKey:kCGImagePropertyOrientation
    ] };
NSArray *adjustments = [myImage autoAdjustmentFiltersWithOptions:
    options];
for (CIFilter *filter in adjustments) {
    [filter setValue:myImage forKey:kCIInputImageKey];
    myImage = filter.outputImage;
}
```

Listing 3.6 Getting auto enhancement filters and applying them to an image [1]

3.6   Learn More About It

These are the essential official Apple documents on Core Image:

- *Core Image Reference Collection* [9]
- *Core Image Programming Guide* [7]
- *Core Image Filter Reference* [5]
- *Core Image Kernel Language Reference* [6]

The *Core Image Explorer*[10] is an interactive iPhone app that showcases several image filters available in Core Image for iOS [16].

Additionally, you may want to check [10] for a detailed, hands-on, interactive Core Image Tutorial in Swift, which includes functionality for getting photos from (and saving filtered images to) the iPhone/iPad camera roll.

For a technical discussion on how to extend Core Image effects to live video, see [4].

Last, but not least, the richly illustrated (e-)book *Core Image for Swift* [15] provides a comprehensive guide to Core Image for iOS Swift developers. The book's companion app, *Filterpedia*,[11] is an open-source iPad app for exploring a broad range of image filters offered by the Core Image framework.

References

1. Auto Enhancing Images. https://apple.co/2VdbyDD. Accessed: 2020-06-23.
2. CIDetector Class Reference. https://apple.co/3hT2TzH. Accessed: 2020-06-23.
3. CIFilter Class Reference. https://apple.co/2YoPHLq. Accessed: 2020-06-23.
4. Core Image and Video. https://bit.ly/2Nofmxm. Accessed: 2020-06-23.
5. Core Image Filter Reference. https://apple.co/3fO0LaM. Accessed: 2020-06-23.
6. Core Image Kernel Language Reference. https://apple.co/2VbrWnV. Accessed: 2020-06-23.

---

[10]https://github.com/objcio/issue-21-core-image-explorer.
[11]https://github.com/FlexMonkey/Filterpedia.

7. Core Image Programming Guide. https://apple.co/37hvKZT. Accessed: 2020-06-09.
8. Core Image Programming Guide—Processing Images. https://apple.co/37UFqJX. Accessed: 2020-06-23.
9. Core Image Reference Collection. https://apple.co/30t8Xst. Accessed: 2020-06-10.
10. Core Image Tutorial: Getting Started. https://bit.ly/2Yq9aLv. Accessed: 2020-06-23.
11. Detecting Faces in an Image. https://apple.co/2BAXlti. Accessed: 2020-06-23.
12. Processing an Image Using Built-in Filters. https://apple.co/3doFqmN. Accessed: 2020-06-23.
13. Querying the System for Filters. https://apple.co/3hU99Y9. Accessed: 2020-06-23.
14. Subclassing CIFilter: Recipes for Custom Effects. https://apple.co/3i1JGw7. Accessed: 2020-06-23.
15. S. J. Gladman. *Core Image for Swift: Advanced Image Processing for iOS*. Reaction Diffusion Limited, 2016.
16. W. Moore. An Introduction to Core Image. https://bit.ly/2VebFii. Accessed: 2020-06-23.

Chapter 4
Machine Learning with Core ML

4.1 Machine Learning: The Basics

Machine learning (ML) is a branch of artificial intelligence (AI), which consists of the design and implementation of algorithms that learn directly from data without relying on a predetermined equation as a model. This is in contrast with rule-based AI systems, in which the rules are usually learned from a human expert and explicitly encoded into the AI solution, a paradigm that was used in the expert systems of the 1980s, for example. Figure 4.1 illustrates the shift from rule-based AI to machine-learning-based AI: rather than having humans input rules (a program) and data to be processed according to these rules, from which intelligent answers can be derived, under the machine learning paradigm, programmers input data as well as the answers expected from the data, and derive rules which can be later applied to new data to produce original answers [27].

A ML system (or *model*) is trained by presenting many examples relevant to a task, which it uses to find statistical structure that eventually allows the model to come up with generalizable rules for automating the task [27]. For instance, if you want to automate the task of classifying your pets' pictures, you could train a machine learning system with many examples of pictures previously labeled by humans (as 'cat' or 'dog', for example), and build a model that learns rules for classifying new (unlabeled) images into 'cat' or 'dog' with a high degree of accuracy.

Machine learning has become the most popular and most successful subfield of AI, in great part due to the availability of fast hardware and very large datasets. ML solutions are being applied to a variety of scientific problems and industries, including: finance, computational biology, energy production, natural language processing, and—most relevant for this book—image processing and computer vision, in tasks such as face and object recognition, pedestrian detection, and medical image analysis, among many others.

© The Author(s) 2020
O. Marques, *Image Processing and Computer Vision in iOS*, SpringerBriefs in
Computer Science, https://doi.org/10.1007/978-3-030-54032-6_4

Fig. 4.1 From rule-based AI
to machine learning AI
(redrawn from [27])

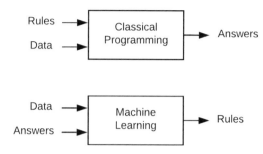

It is said that an algorithm (called an *agent* in AI parlance) is learning if it improves its performance on future tasks after making observations about the world. Such improvements usually come from whatever type of feedback is available to learn from. The types of feedback determine the main types of learning [41]:

• In *unsupervised learning* the ML algorithm learns patterns in the input even though no explicit feedback is supplied.
• In *reinforcement learning* the ML algorithm learns from a series of reinforcements that can act as rewards or punishments.
• In *supervised learning* the ML algorithm observes some labeled examples consisting of input-output pairs and learns a function that maps from input to output.
• In *semi-supervised learning* the ML algorithm is given a few labeled examples and uses the learned knowledge to handle a large collection of unlabeled examples.

The vast majority of ML applications currently in use supervised learning techniques, which can be further divided into two main groups:

1. *Classification* techniques, which consist of building models which are then used to classify input data into categories, for example, in the context of object recognition, classify an input image as a cat or dog.[1]Examples of classification techniques include: naive Bayes, k-nearest neighbors, logistic regression, support vector machine (SVM), decision trees, and neural networks.
2. *Regression* techniques, which consist of building models which are then used to predict a numerical value that varies continuously, for example, given an image of a face, estimate the rotation angles of the face relative to the camera (called *yaw*, *pitch*, and *roll*).[2]Examples of regression techniques include: linear regression, decision trees, and neural networks.

---

[1]For more on the cat vs. dog classification problem in machine learning, check the associated Kaggle challenge [16].

[2]This is known as the head-pose estimation problem; see [38] for a survey.

Classification models predict *discrete* responses whereas regression models predict *continuous* responses.

Selecting which ML algorithm to use for a specific problem can be an overwhelming task! There are dozens of supervised and unsupervised machine learning algorithms in the literature, and each takes a different approach to learning: there is no "best method" or "one size fits all." Finding the right algorithm is often an iterative process where a fair amount of trial and error should be expected. Selecting a ML algorithm also depends on many application-specific factors, such as: the size and type of data available, the insights that the resulting model should get from the data, and how those insights will be used within the context of the application.[3]

4.2   The Era of Deep Learning

During the past few years, the emergence of a machine learning paradigm known as *deep learning* has enabled the development of intelligent systems that can, in many cases, demonstrate better-than-human performance. Deep learning techniques use neural networks with more neurons than previous neural networks and more complex ways of connecting layers. They have the ability to handle two tasks that were previously tackled separately: *feature extraction* (automatically learn which features to extract from the raw data) and *classification* (or *regression*). For computer vision tasks, a specific architecture known as Convolutional Neural Networks (often referred to as CNNs or ConvNets) [30, 37]—designed to handle raw image pixels as input data and, therefore, dispensing with the need to extract image features using hand-crafted algorithms (see Fig. 4.2)—has been extensively used in a variety of image analysis and computer vision tasks. Inspired by the success of using CNNs in image classification challenges [35], deep-learning-based solutions have become ubiquitous in modern computer vision.[4]

The term *deep learning* encompasses an ever-growing list of different types of neural network architectures besides CNNs, among them: autoencoders (and their variants), Recurrent Neural Networks (RNNs), Generative Adversarial Networks (GANs), and many others, which are beyond the scope of this book.

Thanks to a paradigm known as *transfer learning*, it is possible to use pre-trained deep learning models that can be adapted, re-trained (at a fraction of the time and computational cost), and used in another problem, (vaguely) related to the one for which the model was originally conceived. For example, a deep learning architecture trained on the ImageNet dataset [28]—1 million images from 1000 different categories—can be adapted to a specialized visual recognition task, such

---

[3]The scikit-learn algorithm cheat-sheet [8] provides a (highly simplified) flowchart to help beginners navigate the space of candidate ML algorithms for a given problem.

[4]As I mentioned in Chap. 2, it is commonly said that the history of computer vision will be written in two volumes: (1) before deep learning (1950s–2012); and (2) after deep learning (2012–present).

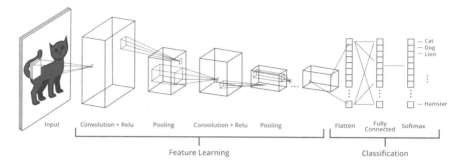

Fully
Connected

Fig. 4.2 A typical CNN allows end-to-end training, taking raw image pixels as input and producing a result (in this case, the predicted label in an image classification task) at the output

as classifying flowers among several different types. In Sect. 4.4 we describe some of the pre-trained models available in Core ML and how they can be used in your iOS app.

4.3   Core ML Basics

Core ML is a machine learning framework introduced by Apple in 2017. The Core ML framework allows developers to integrate machine learning models into their app (Fig. 4.3). In this context, a *trained model* is the result of the training phase in machine learning, where a machine learning algorithm has been exposed to a set of training data and has learned generalizable descriptions of the data, which can be used to make predictions on never-before-seen data. With Core ML you can use existing models or build and train your own (using the Create ML app bundled with Xcode) [9].

Core ML is the foundation for domain-specific frameworks and functionality. It is built on top of low-level primitives like Accelerate and BNNS, as well as Metal Performance Shaders. Core ML supports other Apple frameworks such as: *Vision* for analyzing images, *Natural Language* for processing text, *Speech* for converting audio to text, and *Sound Analysis* for identifying sounds in audio (Fig. 4.4) [9].

Core ML supports the *Vision* framework (Chap. 5) for image analysis and is optimized for on-device performance, thereby minimizing memory footprint and power consumption. The advantages of running machine learning applications strictly on the device include: the ability to retrain or fine-tune your models on-device, with that user's data; ensuring the privacy of user data; and guaranteeing that the app remains functional and responsive even in the absence of a network connection [9]. Additionally, the Core ML API [21] enables developers to build custom workflows and handle advanced use cases.

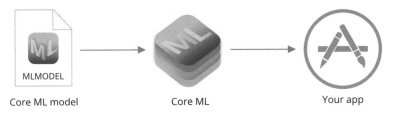

Core ML model   Core ML   Your app

Fig. 4.3   Core ML (redrawn from [9])

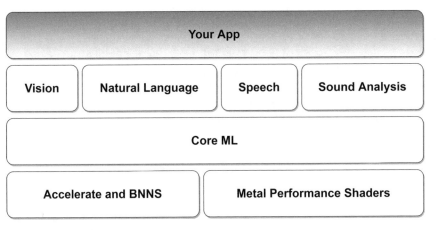

Fig. 4.4   Core ML stack (redrawn from [9])

Core ML 2 was released in 2018 and showed substantial improvements in terms of speed (Apple claims it is 30% faster than the previous version due to batch prediction) and size of the AI models, which can be greatly reduced thanks to quantization [4]. Core ML 2 can update models from cloud services such as Amazon Web Services (AWS) or Microsoft Azure at runtime, and it includes a converter that works with popular deep learning frameworks and libraries such as Facebook's Caffe and Caffe2, Keras, scikit-learn, and Google's TensorFlow Lite [4].

Core ML 3—introduced during the WWDC 2019 conference—allows you to fine-tune an updatable model[5]on the device during runtime, which enables the creation of personalized experiences for each user, a concept used by Apple itself in its Face ID: Initially, Apple ships a model down to the device that recognizes generic faces; during Face ID set up, each user can fine-tune the model to recognize their face, using data that should not leave the device [10].

---

[5]An updatable model is a Core ML model that is marked as updatable. To learn more about how to modify an updatable Core ML model by running an update task with labeled data, see [19].

The latest updates on Core ML,[6]announced during the WWDC 2020 conference include: new layer types, support for encrypting models, and the ability to host model updates on CloudKit, which lets you update your models independently of the app [3].

4.4 Using Pre-trained Models with Core ML

Core ML supports several machine learning models, including many types of neural networks, tree ensembles, support vector machines, and generalized linear models. These models must be saved in the Core ML model format (with a `.mlmodel` file extension), an open file format that describes the layers in the model, its input and outputs, the class labels, all the learned parameters (weights and biases), and any preprocessing that needs to happen on the data [32].

Examples of models that can be downloaded directly from Apple's website [11] include: image classification (using MobileNetv2 [42], ResNet50 [31], or SqueezeNet [33]), image segmentation using DeeplabV3 [26], pose estimation [39], depth estimation [36], handwritten digit classification, drawing classification, and object detection using YOLOv3 [40]. See [5] and [6] for a (growing) list of currently available third-party Core ML models.

Moreover, third-party models made publicly available by research groups and universities, which may not be in the Core ML model format, can be converted using Core ML Tools [12]—an open-source Python package used to create models in the Core ML format, if your model is created and trained using a supported third-party machine learning framework (e.g. TensorFlow, PyTorch, etc).

Integrating a Core ML model into an app is an easy and straightforward process [22]. It consists essentially of:

1. Loading a pre-trained model.
2. Making predictions using new data.
3. Use the results of the predictions in a creative/profitable way.

4.5 Training Your Own Models with Create ML

Since the release of Core ML 2, the framework also supports the ability to create machine learning models to use in your app using Create ML; the trained model can eventually be saved (with a `.mlmodel` file extension) and added to an existing Core ML enabled app [13].

---

[6]The version numbering of Core ML became rather confusing in recent years; when you see a reference to Core ML (without a number) it might be referring to the latest version [3].

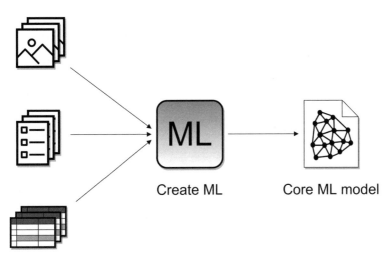

Create ML   Core ML model

Fig. 4.5 Using Create ML to learn a Core ML model from different types of data (redrawn from [13])

Create ML allows building and training Core ML models on the Mac with no code. The easy-to-use app interface and models available for training make the process easier than ever, so all you need to get started is your training data. You can even take control of the training process with features like snapshots and previewing to help you visualize model training and accuracy [23]. Using Create ML and your own data, you can train custom models to perform tasks such as recognizing images, extracting meaning from text, or finding relationships between numerical values (Fig. 4.5). Models trained using Create ML are in the Core ML model format and are ready to use in your app [13].

Create ML enables the implementation of the classical ML workflow (Fig. 4.6), where the model is trained by being shown representative samples of labeled data from the training set, evaluated using the validation set, and refined after multiple iterations and adjustments to the model's (hyper)parameters. Once the model is performing well enough, it is considered ready to be integrate it into your app using Core ML [13].

A popular alternative to Create ML is Turi Create, an open source Python library containing a collection of task-based tools for training machine-learning models, manipulating data, and exporting models in the CoreML model format. Turi Create runs on anything that supports Python, and is maintained and supported by Apple [29].

Fig. 4.6 Classical ML
workflow using Create ML
(redrawn from [13])

4.6   Computer Vision and Core ML

Core ML makes it easier to use some of the best performing contemporary deep
learning models in mobile computer vision apps. You can combine the Vision framework (Chap. 5) with Core ML to perform tasks such as image classification [20].

Essentially, the Core ML framework enables the use of pre-trained machine
learning models to classify input data, whereas the Vision framework preprocesses
those images and works with Core ML to apply those classification models to
images.

To set up a Vision request using a pre-selected model, you must create an instance
of that class and use its `model` property to create a `VNCoreMLRequest` object.
You can then use the request object's completion handler to specify a method
to receive results from the model after you run the request. Note that you can
use the request's `imageCropAndScaleOption` property to handle a common
preprocessing step by which you can match the image layout to one expected by the
model [20] (Listing 4.1).

```
1 // load model
2 let model = try VNCoreMLModel(for: MobileNet().model)
3
4 // set up request
5 let request = VNCoreMLRequest(model: model, completionHandler: {[
    weak self] request, error in
6     self?.processClassifications(for: request, error: error)
7 })
8 request.imageCropAndScaleOption = .centerCrop
9 return request
```

Listing 4.1 Typical code skeleton for setting up an image analysis request that uses a Core ML
model to process images [20]

You can then run the Vision request (Listing 4.2). Note that the image's
orientation is passed to the image request handler to ensure proper handling of input
images with arbitrary orientations [20].

```
DispatchQueue.global(qos: .userInitiated).async {
    let handler = VNImageRequestHandler(ciImage: ciImage,
    orientation: orientation)
    do {
        try handler.perform([self.classificationRequest])
    } catch {
        print("Failed to perform classification.\n\(error.
localizedDescription)")
    }
}
```

Listing 4.2 Typical code skeleton for running an image classification request [20]

Finally, if the Vision request succeeded, its `results` property contains
`VNClassificationObservation` objects describing possible classifications
identified by the ML model [20] (Listing 4.3).

The functionality of the `VNCoreMLRequest` object extends beyond predicting
a single feature, such as the image classification example above. It also handles:

- Image analysis scenarios in which the selected Core ML model's outputs include
  at least one output whose feature type is *image*; in those cases, the Vision
  framework treats that model as an image-to-image model and stores the results
  in a `VNPixelBufferObservation` object, which correspond to the output
  image produced by a Core ML image analysis request [24].
- Prediction applications (rather than classification or image-to-image processing),
  in which the Core ML model predicts multiple features; in those cases, the Vision
  framework stores the results in a `VNCoreMLFeatureValueObservation`
  object, essentially a collection of key-value pairs [24].

```
func processClassifications(for request: VNRequest, error: Error
    ?) {
    DispatchQueue.main.async {
        guard let results = request.results else {
            self.classificationLabel.text = "Unable to classify
image.\n\(error!.localizedDescription)"
            return
        }
        // The 'results' will always be '
VNClassificationObservation's, as specified by the Core ML
model in this project.
        let classifications = results as! [
VNClassificationObservation]
```

Listing 4.3 Typical code skeleton for processing the results of an image classification request [20]

4.7 Learn More About It

There are numerous books, journals, and online resources where you can learn more about machine learning and deep learning. A curated list of machine learning books by Jason Brownlee [25] is an excellent starting point. You should also check this list of machine learning and deep learning frameworks, libraries, books, blogs, courses, and software on GitHub [7].

The book by Geldard et al. [29] provides an excellent, practical, and up-to-date coverage of AI, machine learning, and deep learning with focus on iOS app development. A must read!

The book by Pete Warden [45] teaches how to build mobile apps with the popular deep learning framework, TensorFlow.

Apple's Machine Learning site [17] has links to papers and other resources that might be of interest to readers who want to dive into deeper research topics.

A good starting point to explore Core ML is the official Apple "Machine Learning—Apple Developer" portal [23], which contains links to examples, videos, and the official documentation.

Matthijs Hollemans maintains an excellent blog on machine learning and deep learning, with focus on mobile app development, iOS, and Core ML [18]. In his blog, you will find detailed technical descriptions of the latest updates in the Apple machine learning ecosystem and a variety of other relevant topics for iOS app developers who want to build intelligent apps.

Other good tutorials and examples (with associated source code) include:

- "Creating an Image Classifier Model," an Apple step-by-step tutorial on how to train a machine learning model to classify images, and add it to your Core ML app [15].
- "Core ML and Vision Tutorial: On-device training on iOS," by Christine Abernathy and Audrey Tam [10]: a step-by-step tutorial on how to fine-tune a model on the device using Core ML and Vision Framework.
- "Beginning Machine Learning with Keras and Core ML," by Audrey Tam: a detailed tutorial that teaches how to train a CNN model in Keras, convert it to Core ML, and integrate it into an iOS app [44].
- "Introduction to Create ML: How to Train Your Own Machine Learning Model in Xcode 10," by Sai Kambampati [14], provides a few good introductory hands-on examples of the Create ML workflow for training and evaluating your own models.
- "A simple demo for Core ML in Swift [1]," companion code for the article "Introduction to Core ML: Building a Simple Image Recognition App", by Sai Kambampati [34], which uses the Inception v3 model and shows how to build an image recognition demo app with camera and photo library support.
- "A simple game developed using Core ML [2]," companion code for the article "Creating a Simple Game With Core ML in Swift 4", by Mitchell Sweet [43], which shows how to create a simple yet clever scavenger hunt game in which the player must use their iPhone camera to find an example of an object whose name

appears at the top of the screen. Once the player finds the object and points their phone at it, the app—using a machine learning algorithm built upon a pre-trained Inception V3 model—will recognize the object, add a point to the player's score, and then ask for another one, repeating this process until the time is up.

References

1. A simple demo for Core ML in Swift. https://bit.ly/2YrQMlx. Accessed: 2020-06-23.
2. A Simple Game developed using Core ML. https://bit.ly/3fUICs2. Accessed: 2020-06-23.
3. Apple machine learning in 2020: What's new? https://bit.ly/2DPY5vu. (Accessed on 2020-07-18).
4. Apple's Core ML 2 vs. Google's ML Kit: What's the difference? https://bit.ly/3dsvx7x. Accessed: 2020-06-23.
5. Awesome Core ML Models. https://bit.ly/316FDsh. Accessed: 2020-06-23.
6. Awesome Core ML Models. https://bit.ly/2VajOUX. Accessed: 2020-06-23.
7. Awesome Machine Learning. https://bit.ly/31aYJxa. Accessed: 2020-06-23.
8. Choosing the right estimator—scikit-learn documentation. https://bit.ly/2Oh3u0B. Accessed: 2020-07-11.
9. Core ML. https://apple.co/316Pf6h. Accessed: 2020-06-23.
10. Core ML and Vision Tutorial: On-device training on iOS. https://bit.ly/37UaA4c. Accessed: 2020-06-23.
11. Core ML Models—Apple Developer. https://apple.co/3gJiY9Q. Accessed: 2020-07-11.
12. Core ML Tools. https://coremltools.readme.io/docs. (Accessed on 2020-07-17).
13. Create ML. https://apple.co/2VekAQy. Accessed: 2020-06-23.
14. Create ML: How to Train Your Own Machine Learning Model in Xcode 10. https://bit.ly/3eC4V4p. (Accessed on 2020-07-18).
15. Creating an Image Classifier Model. https://apple.co/2B6ExBY. Accessed: 2020-06-23.
16. Dogs vs. Cats—Kaggle. https://bit.ly/2V9EhJe. Accessed: 2020-06-23.
17. Machine Learning Research at Apple. https://bit.ly/mlresearchapple. Accessed: 2020-06-23.
18. Matthijs Hollemans—Blog. https://machinethink.net/blog/. Accessed: 2020-07-11.
19. Personalizing a model with on-device updates. https://apple.co/2WxIYxb. (Accessed on 2020-07-18).
20. Classifying Images with Vision and Core ML. https://apple.co/3fRVpeK, 2017. Accessed: 2020-06-23.
21. Core ML API. https://apple.co/37XzZK8, 2018. Accessed: 2020-06-23.
22. Integrating a Core ML Model into Your App. https://apple.co/2V8HVD9, 2018. Accessed: 2020-06-23.
23. Machine Learning—Apple Developer. https://apple.co/3dniNyY, 2018. Accessed: 2020-06-23.
24. VNCoreMLRequest: an image analysis request that uses a Core ML model to process images. https://apple.co/37WAmVw, 2018. Accessed: 2020-06-23.
25. J. Brownlee. Machine learning books. https://bit.ly/2B4nksT. Accessed: 2020-06-23.
26. L.-C. Chen, G. Papandreou, I. Kokkinos, K. Murphy, and A. L. Yuille. DeepLab: Semantic Image Segmentation with Deep Convolutional Nets, Atrous Convolution, and Fully Connected CRFs, 2016.
27. F. Chollet. *Deep learning with Python*. Manning Publications Co, Shelter Island, NY, 2018.
28. J. Deng, W. Dong, R. Socher, L.-J. Li, K. Li, and L. Fei-Fei. ImageNet: A Large-Scale Hierarchical Image Database. In *CVPR09*, 2009.
29. M. Geldard. *Practical artificial intelligence with Swift : from fundamental theory to development of AI-driven apps*. O'Reilly Media, Sebastopol, CA, 2019.

30. I. Goodfellow, Y. Bengio, and A. Courville. *Deep learning*. MIT press, 2016.
31. K. He, X. Zhang, S. Ren, and J. Sun. Deep residual learning for image recognition. *CoRR*, abs/1512.03385, 2015.
32. M. Hollemans. iOS 11: Machine Learning for everyone. https://bit.ly/2YpyiCm. Accessed: 2020-06-23.
33. F. N. Iandola, S. Han, M. W. Moskewicz, K. Ashraf, W. J. Dally, and K. Keutzer. Squeezenet: AlexNet-level accuracy with 50x fewer parameters and <0.5 MB model size. *arXiv:1602.07360*, 2016.
34. S. Kambampati. Introduction to Core ML: Building a Simple Image Recognition App. https://bit.ly/319FYdF. Accessed: 2020-06-23.
35. A. Krizhevsky, I. Sutskever, and G. E. Hinton. Imagenet classification with deep convolutional neural networks. In *Advances in neural information processing systems*, pages 1097–1105, 2012.
36. I. Laina, C. Rupprecht, V. Belagiannis, F. Tombari, and N. Navab. Deeper depth prediction with fully convolutional residual networks, 2016.
37. Y. LeCun, Y. Bengio, and G. Hinton. Deep learning. *Nature*, 521(7553):436–444, 2015.
38. E. Murphy-Chutorian and M. M. Trivedi. Head pose estimation in computer vision: a survey. *IEEE Transactions on Pattern Analysis and Machine Intelligence*, 31(4):607–626, 2009.
39. G. Papandreou, T. Zhu, N. Kanazawa, A. Toshev, J. Tompson, C. Bregler, and K. Murphy. Towards Accurate Multi-person Pose Estimation in the Wild, 2017.
40. J. Redmon and A. Farhadi. YOLOv3: An Incremental Improvement, 2018.
41. S. Russell and P. Norvig. *Artificial Intelligence: A Modern Approach*. Prentice Hall Series in Artificial Intelligence. Prentice Hall, 2010.
42. M. Sandler, A. Howard, M. Zhu, A. Zhmoginov, and L.-C. Chen. Mobilenetv2: Inverted residuals and linear bottlenecks, 2018.
43. M. Sweet. Creating a Simple Game With Core ML in Swift 4. https://bit.ly/31bk0ak. Accessed: 2020-06-23.
44. A. Tam. Beginning Machine Learning with Keras & Core ML. https://bit.ly/2YoUKvk. Accessed: 2020-06-23.
45. P. Warden. *Building Mobile Applications with TensorFlow*. O'Reilly.

Chapter 5
Computer Vision and Image Analysis with the Vision Framework

5.1 Introduction

The Vision Framework [22] was introduced with iOS 11 in 2017. It was conceived to be a one-stop solution for IPCV problems. It has a simple and consistent interface; runs of iOS, macOS, and tvOS; and is privacy-oriented [27].

The Vision Framework encapsulates advanced capabilities for handling face detection and recognition, barcode and QR code detection, text detection, horizon detection, image registration (alignment), as well as object detection and tracking. It also interfaces with Core ML (Chap. 4) models for image analysis and machine learning tasks in the same workflow.

The Vision Framework is another example of turning computer vision into a commodity and making sophisticated computer vision capabilities available to software developers who might not have a strong computer vision/image processing/machine learning background.

The main motivating factors for empowering sophisticated computer vision tasks on the device (rather than over the cloud) are: privacy (images and videos remain on the user's device), cost (no usage fees or data transfer costs), and performance (no latency, fast execution) [28].

5.2 The Image Analysis Pipeline

The Vision Framework breaks down a IPCV problem into three components [27]:

- **What to Process?**
  This is implemented using a family of requests, whose abstract superclass is
  VNRequest.

© The Author(s) 2020
O. Marques, *Image Processing and Computer Vision in iOS*, SpringerBriefs in
Computer Science, https://doi.org/10.1007/978-3-030-54032-6_5

- **How to Process?**
  This is implemented using request handlers, or engines, such as
  VNImageRequestHandler for images and
  VNSequenceRequestHandler for image sequences (i.e., videos).
- **Where to Look for Results?**
  This is implemented using a family of observations, whose abstract superclass is
  VNObservation.

The Vision Framework adopts a three-stage pipeline in which everything starts with a *Request*, which is handled by a *RequestHandler*, that in turn produces results, known as *Observations*, which are task-dependent: an observation could be a label (as a result of a classification process), or the bounding boxes of detected objects (in the case of object detection tasks), or the trajectory of an object tracked across a video sequence.

Request handlers come in two types [22]:

- Image request handler (VNImageRequestHandler): an object that processes one or more image analysis requests pertaining to a single image. The VNImageRequestHandler object can be used for interactive exploration of an image; consequently, it holds on to the image for its lifecycle and allows optimization of various requests performed on an image [28].
- Sequence request handler (VNSequenceRequestHandler): an object that processes image analysis requests pertaining to a sequence of multiple images, i.e., a video. The VNSequenceRequestHandler object does not optimize for multiple requests on an image [28].

For the image request handler, the three steps (create request, handle request, produce results) are usually encapsulated in code in the form illustrated by the skeleton in Listing 5.1. In the case of the sequence request handler, the equivalent steps are shown in Listing 5.2.

```
1  // Create request
2  let faceDetectionRequest = VNDetectFaceRectanglesRequest()
3
4  // Create request handler
5  let myRequestHandler = VNImageRequestHandler (url: fileURL,
       options: [:])
6
7  // send the requests to the request handler
8  myRequestHandler.perform([faceDetectionRequest])
9
10 // Do we have a face?
11 for observation in faceDetectionRequest.results as! [
       VNFaceObservation] {
12     // do something
13 }
```

Listing 5.1 Typical code skeleton for handling a request on an image for the case of face detection [28]

```
// Create sequence request handler
let myRequestHandler = VNSequenceRequestHandler()

// Start tracking with an observation
let observations = detectionRequest.results as! [
    VNDetectedObjectObservation]
let objectsToTrack = observations.map { VNTrackObjectRequest(
    detectedObjectObservation: $0) }

// Run the requests
requestHandler.perform(objectsToTrack, on: pixelBuffer)

// Inspect results
for request in objectsToTrack
  for observation in request.results as! [
    VNDetectedObjectObservation]
```

Listing 5.2 Typical code skeleton for handling requests on image sequences (videos) for the case of object tracking [28]

5.3 Practical Recommendations

Image Types

The choice of image type to use depends on other aspects of your application, for example, where the images come from. You are not required to pre-scale the image, but you should pass the EXIF orientation of the image as a parameter. Vision Framework supports the following image types [22, 28]:

- CVPixelBufferRef: a reference to a Core Video pixel buffer object, i.e., the image that comes from a CMSampleBuffer in the VideoDataOut of a camera stream.
- CGImageRef: used if the image was already used in the UI; both UIImage and NSImage have accessors for CGImageRefs.
- CIImage: used when you are already using Core Image in your application.
- NSURL: used to inform the URL for image files on disk.
- NSData: used for images from the Web.

Image Orientation

Not all algorithms are orientation-agnostic and the images you use in your app might not be upright. You should get the precise definition of what is upright by looking into the EXIF orientation metadata. When using an image URL as input, the Vision framework reads the EXIF orientation from the file; when capturing from a live feed, the orientation has to be inferred from `UIDevice.current.orientation` and it needs to be mapped to a `CGImagePropertyOrientation` [25].

Coordinate System

The coordinate system used in the Vision framework has its origin at the lower-left corner and assumes that all processing is in relation to the image in upright coordinates. It uses normalized coordinates (0.0 to 1.0). In the case of face detection, facial landmarks' coordinates are relative to the face rectangle, which can be converted into image coordinates by using conversion utils such as `VNImageRectForNormalizedRect` (from `VNUtils.h`) [25].

Performance Aspects

Vision tasks can be time consuming and processing intensive. It is recommended that you dispatch your work on a queue with appropriate quality of service (QoS) and use the completion handler to process the results [28].

Face Detection

Vision Framework offers a robust face detection module, capable of detecting faces that are small, partially occluded, or in profile (rather than frontal) view, as well as handling artifacts such as glasses and hats appropriately. It also extracts the facial landmarks (corners of mouth, centers of eyes, etc.).

The face detection functionality provided by the Vision Framework is more accurate than the equivalent functionality provided by Core Image (see Sect. 3.4)[1] and `AVFoundation`, but it lags behind both alternatives in terms of processing time and power usage [28].

5.4   Integration with Core ML

The Vision Framework allows you to use a pre-trained machine learning model for tasks such as image classification, character recognition, and object recognition.

---

[1]Note that the `CIDetector` will remain as is in Core Image, whereas new algorithms for improved face detection are expected to became available through Vision Framework.

The workflow is typically structured in three main parts [21]:

1. **Set up Vision with a Core ML Model**
   Core ML automatically generates a Swift class that provides easy access to your
   ML model (e.g., `MobileNet`). To set up a Vision request using that model,
   you must create an instance of that class and use its model property to create a
   `VNCoreMLRequest` object. You can then use the request object's completion
   handler to specify a method to receive results from the model after having run
   the request.

   Note that several ML models expect the input images to be of a certain fixed
   size and aspect ratio; therefore, it is common to have a pre-processing step (using
   Core Image) to center, scale and/or crop each image to fit the expected size. In
   this case, centering and cropping can be achieved elegantly in a single line of
   code, by setting the request's `imageCropAndScaleOption` property (using
   the `.centerCrop` option) to match the image layout the model was trained
   with (Listing 5.3).

2. **Run the Vision Request**
   You can then create a `VNImageRequestHandler` object with the image to
   be processed, and pass the requests to that object's `perform()` method. This
   method runs synchronously, i.e., you must use a background queue so that the
   main queue isn't blocked while your requests execute[2] (Listing 5.4).

3. **Handle Image Classification Results**
   Once the request has been processed, the Vision request's completion handler
   should indicate whether the request succeeded or resulted in an error. In the case
   of success, the `results` property of the `VNCoreMLRequest` object should
   contain one or more `VNClassificationObservation` objects describing
   possible classifications identified by the ML model (Listing 5.5).

```
let model = try VNCoreMLModel(for: MobileNet().model)

let request = VNCoreMLRequest(model: model, completionHandler: {
    [weak self] request, error in
    self?.processClassifications(for: request, error: error)
})
request.imageCropAndScaleOption = .centerCrop
return request
```

Listing 5.3 Setting up Vision with a Core ML model [21]

---

[2]To ensure proper handling of input images with arbitrary orientations, you should also pass the
image's orientation to the image request handler.

```
DispatchQueue.global(qos: .userInitiated).async {
    let handler = VNImageRequestHandler(ciImage: ciImage,
    orientation: orientation)
    do {
        try handler.perform([self.classificationRequest])
    } catch {
        /*
        This handler catches general image processing errors.
        The `classificationRequest`'s completion handler
        `processClassifications(_:error:)` catches errors
        specific to processing that request.
        */
        print("Failed to perform classification.\n\(error.
        localizedDescription)")
    }
}
```

Listing 5.4 Running the Vision request [21]

```
func processClassifications(for request: VNRequest, error: Error
    ?) {
    DispatchQueue.main.async {
        guard let results = request.results else {
            self.classificationLabel.text = "Unable to classify
            image.\n\(error!.localizedDescription)"
            return
        }
        /*
        The 'results' will always be
        `VNClassificationObservation`'s, as
        specified by the Core ML model in
        this project.
        */
        let classifications = results as! [
        VNClassificationObservation]
```

Listing 5.5 Handling image classification results [21]

5.5   Examples of IPCV Solutions Using Vision Framework

In this section we provide pointers to selected examples/demos from Apple's official site that cover standard IPCV tasks such as: object detection and recognition, image classification and object tracking, as well as more sophisticated applications such as: pose detection and human activity recognition. All examples (and their source code) are freely available online. Some of these examples have been mentioned in WWDC presentations whose video recordings are also freely available.

Classifying Images with Vision and Core ML

This is a straightforward example of how to implement an image classification pipeline using the Vision framework to preprocess photos and a pretrained Core ML model to classify them [21].

Detecting Objects in Still Images

This is an example of how to use the Vision framework to demarcate rectangles, faces, barcodes, and text in images [3].

Recognizing Objects in Live Capture

This example teaches how to use Vision framework algorithms to perform object recognition in live capture, showing step by step how to set up your camera for live capture, incorporate a Core ML model into Vision, and parse results as classified objects. [10].

Training a Create ML Model to Classify Flowers

This is an excellent example that shows how to train a flower classifier using Create ML in Swift Playgrounds Sect. 4.5, and apply the resulting model to real-time image classification using the Vision framework [17]. It uses the same code as the robot shop demo in the WWDC 2018 session *Vision with Core ML* [25].

Tracking Multiple Objects or Rectangles in Video

This example shows how to apply Vision algorithms to track objects or rectangles throughout a video, which consists of: (1) picking an initial object to track; (2) creating Vision tracking requests to follow that object; and (3) parsing results from the object or rectangle tracker [15].

Tracking the User's Face in Real Time

This example shows how to create requests to track human faces and interpret the results of those requests in real time [16].

Highlighting Areas of Interest in an Image Using Saliency

This example [6] shows how to quantify and visualize where people are likely to look in an image, using the Vision Framework's ability to compute two types of saliency within an image: (1) object-based saliency, which highlights foreground objects and provides a coarse segmentation of the main subjects in an image and attention-based; and (2) attention-based saliency, which highlights what people are likely to look at [2].

Building a Feature-Rich App for Sports Analysis

This example highlights some of the latest (iOS 14+) features of the Vision Framework to detect and classify human activity in real time using computer vision and machine learning [1].

Text Recognition Examples

These examples highlight the text recognition capabilities added to the Vision Framework as of iOS 13.0—the VNRecognizeTextRequest image analysis request that finds and recognizes text in an image and the VNRecognizedText Observation request that detects and recognizes regions of text in an image [11].

- **Locating and Displaying Recognized Text on a Document**: shows how to overlay text recognition output from the document scanner onto an image, reporting progress throughout [7].
- **Reading Phone Numbers in Real Time**: shows how to analyze and filter phone numbers from text recognized in live capture, building evidence over time [9].
- **Structuring Recognized Text on a Document**: shows how to detect, recognize, and structure text on a business card or receipt using Vision and VisionKit [12].

6 Learn More About It

A good starting point to explore the Vision Framework is the "Image Classification with Vision and Core ML" example [23]. Other good tutorials and examples (with associated source code) include:

- Getting Started with Vision on iOS 11, by Jeffrey Bergier: a step-by-step tutorial to perform object tracking using the Vision Framework [24].
- HelloVision [5], companion code for the article "Swift World: What's new in iOS 11–Vision" [13], which covers machine learning image analysis, text detection, face (and face landmarks) detection, object tracking, and barcode detection.
- Vision Framework Demo on Text Detection [19], companion code for the article "Using Vision Framework for Text Detection in iOS 11", by Sai Kambampati [26], which covers text detection in great detail.

These are some recommended videos from the Apple Worldwide Developers Conference (WWDC) (in reverse chronological order):

- 2020: Explore Computer Vision APIs [4]
- 2019: Text Recognition in Vision Framework [14]
- 2019: Understanding Images in Vision Framework [18]
- 2018: Vision with Core ML [20]
- 2018: Object Tracking in Vision [8]

References

1. Building a feature-rich app for sports analysis. https://apple.co/3ezSpCF. Accessed: 2020-07-16.
2. Cropping images using saliency. https://apple.co/30hGbcz. Accessed: 2020-07-16.
3. Detecting objects in still images. https://apple.co/2Wu2FGq. Accessed: 2020-07-16.
4. Explore Computer Vision APIs. https://apple.co/2WqzTq5. Accessed: 2020-07-16.
5. HelloVision: source code for article "Swift World: What's new in iOS 11–Vision". https://bit.ly/2A2vHF6. Accessed: 2020-06-23.
6. Highlighting areas of interest in an image using saliency. https://apple.co/2CiiAk9. Accessed: 2020-07-16.
7. Locating and displaying recognized text on a document. https://apple.co/32oDBnL. Accessed: 2020-07-16.
8. Object Tracking in Vision. https://apple.co/2WpOeDh. Accessed: 2020-07-16.
9. Reading phone numbers in real time. https://apple.co/2OteDvx. Accessed: 2020-07-16.
10. Recognizing objects in live capture. https://apple.co/2Osf3Ce. Accessed: 2020-07-16.
11. Recognizing text in images. https://apple.co/2Oyhnrv. Accessed: 2020-07-16.
12. Structuring recognized text on a document. https://apple.co/2Cl437b. Accessed: 2020-07-16.
13. Swift World: What's new in iOS 11–Vision. https://bit.ly/2NnKVr5. Accessed: 2020-06-23.
14. Text Recognition in Vision Framework. https://apple.co/2OJAK0Z. Accessed: 2020-07-16.
15. Tracking Multiple Objects or Rectangles in Video. https://apple.co/2WqQM3O. Accessed: 2020-07-16.
16. Tracking the User's Face in Real Time. https://apple.co/3eACm7t. Accessed: 2020-07-16.

17. Training a Create ML Model to Classify Flowers. https://apple.co/396V8SP. Accessed: 2020-07-16.
18. Understanding Images in Vision Framework. https://apple.co/3fDqTW5. Accessed: 2020-07-16.
19. Vision Framework Demo on Text Detection. https://bit.ly/2zWz4gA. Accessed: 2020-06-23.
20. Vision with Core ML. https://apple.co/2CaBDwV. Accessed: 2020-07-16.
21. Classifying Images with Vision and Core ML. https://apple.co/3fRVpeK, 2017. Accessed: 2020-06-23.
22. Apple Vision Framework. https://apple.co/37V2bxg, 2018. Accessed: 2020-06-23.
23. Image Classification with Vision and CoreML: sample code (zip). https://apple.co/37VVWcT, 2018. Accessed: 2020-06-23.
24. J. Bergier. Blog: Getting Started with Vision. https://bit.ly/2V7HmcU. Accessed: 2020-06-23.
25. F. Doepke. Vision with CoreML (WWDC 2018 slides). https://apple.co/2WpYsDC. Accessed: 2020-07-16.
26. S. Kambampati. Using Vision Framework for Text Detection in iOS 11. https://bit.ly/2BtBgN3. Accessed: 2020-06-23.
27. S. Kamensky. Object Tracking in Vision (WWDC 2018 slides). https://apple.co/3eB9pbq. Accessed: 2020-07-16.
28. B. Keating and F. Doepke. Vision Framework: building on CoreML (WWDC 2017 slides). https://apple.co/3ett5is, 2018. Accessed: 2020-06-23.

Chapter 6
Opencv and iOS

6.1 Opencv: Overview

Opencv (Open Source Computer Vision Library) is a free, open source computer vision library with thousands of useful algorithms. Written in optimized C/C++, Opencv has C++, C, Python, MATLAB and Java interfaces and supports Windows, Linux, Mac OS, iOS and Android. Opencv was designed for computational efficiency and with a strong focus on real-time applications. The latest version of Opencv is 4.3.0 [2].

The Opencv library contains more than 2500 optimized algorithms that cover many areas in computer vision, from medical imaging, to camera calibration, to stereo vision and robotics. The library has been downloaded more than 18 million times, is extensively used around the world, and contains a large (more than 47,000 users) and active user group [1]. During the past few years, Opencv seems to have achieved one of its main goals, namely "to provide a simple-to-use computer vision infrastructure that helps people build fairly sophisticated vision applications quickly" [14].

Opencv is structured into several modules, which are the primary organizational structure in the library: every function in the library is part of one module. The exact module list has changed over time; these are the currently available modules [5, 14]:

- `core` (core functionality module): contains all of the basic object types and their basic operations;
- `imgproc` (image processing module): contains basic transformations on images, including filters and other convolutional operators, morphological image processing, geometrical image transformations, color space conversions, structural analysis and shape descriptors, feature detection, object detection and tracking, among others;

© The Author(s) 2020
O. Marques, *Image Processing and Computer Vision in iOS*, SpringerBriefs in
Computer Science, https://doi.org/10.1007/978-3-030-54032-6_6

- imgcodecs (image file reading and writing module): contains functions to read and write image files in a large number of file formats—and their variants;
- videoio (video I/O module): contains functions to read and write video streams or image sequences;
- highgui (high-level GUI module): a very lightweight window user interface toolkit, with functions for displaying images and taking simple user input;
- video (video analysis module): contains functions for motion analysis, background subtraction, optical flow estimation, object tracking, and Kalman filtering, among others;
- calib3d (camera calibration and 3D reconstruction module): contains implementations of algorithms to calibrate single cameras as well as stereo or multicamera arrays;
- features2d (2D features framework module): contains algorithms for detecting, describing, and matching keypoint features;
- objdetect (object detection module): contains algorithms for detecting specific objects, such as faces or pedestrians, as well as trainable object detectors.
- dnn (deep neural network module): contains supporting functions for building deep-learning-based computer vision solutions;
- ml (machine learning module): contains a multitude of machine learning algorithms implemented in a way that is compatible with the natural data structures of OpenCV (Sect. 6.2);
- flann (clustering and search in multidimensional spaces module): contains methods to interface with the FLANN (Fast Library for Approximate Nearest Neighbors) library, which contains a collection of algorithms optimized for fast nearest neighbor search in large datasets and for high dimensional features;
- photo (computational photography module): contains useful tools for computational photography, such as denoising, inpainting, high dynamic range (HDR) imaging, and contrast preserving decolorization, among others;
- stitching (images stitching module): contains functions to implement the *registration* and *compositing* portions of an image stitching pipeline, including autocalibration, image warping, features finding and image matching, and exposure compensation, among others;
- gapi (Graph API module): a new OpenCV module that acts as a framework and implements a new graph-based model of execution, targeted to make regular image processing fast and portable.

Starting with OpenCV 3.0, the library has been divided into two parts [14]:

1. mature opencv (whose modules are listed above), which is maintained by the core OpenCV team and contains (mostly) stable code;
2. opencv_contrib, which comprises the current state of the art in computer vision research. It is maintained and developed mostly by the community, may have parts under non-OpenCV license, and may include patented algorithms. Currently the opencv_contrib repository contains modules for specialized deep neural networks, face recognition, text detection and recognition, biologi-

cally inspired vision, advanced image processing and computational photography algorithms, and modern object-tracking algorithms, among others.

OpenCV constitutes one of the most prominent and successful examples of making sophisticated computer vision algorithms available as "commodities" to software developers who want to embed rich image processing and analysis capabilities into their solutions.

6.2 OpenCV: Fundamental Classes and Operations

OpenCV contains hundreds of classes and data structures ranging from basic data types to the larger structures that are used to handle arrays such as images and large matrices and a vast collection of functions that allow us to manipulate this data in multiple useful ways [14].

cv::Mat is the core data structure used to represent any multidimensional matrix in OpenCV. Images in OpenCV are also represented by a cv::Mat, which means that the overwhelming majority of functions in the OpenCV library are members of the cv::Mat class, take a cv::Mat as an argument, or return cv::Mat as a return value; many of them are or do all three [14].

The cv::Mat class stores array data in what can be thought of as an n-dimensional analog of "raster scan order", e.g. in a two-dimensional array, the data is organized into rows, and each row appears one after the other [14].

An instance of cv::Mat acts as a header for the image data and contains information to specify the image format. The image data itself is only referenced and can be shared by multiple cv::Mat instances [11]. Each matrix contains a flags element signaling the contents of the array, a dims element indicating the number of dimensions, rows and cols elements indicating the number of rows and columns, a data pointer to where the array data is stored, and a refcount reference counter, which allows cv::Mat to behave very much like a smart pointer for the data contained in data. The memory layout in data is described by the array step[] [14].

Using the values contained in the step[] array, each pixel of the image can be addressed using pointer arithmetic [11]:

```
uchar *pixelPtr = cvMat.data + rowIndex * cvMat.step[0] +
    colIndex * cvMat.step[1]
```

cv::Algorithm is an abstract base class for most complex algorithms implemented in OpenCV. It provides an API comparable to CIFilter in Apple's Core Image framework (see Chap. 3), where you can create an Algorithm by calling Algorithm::create() with the name of the algorithm, and can set and get various parameters using get() and set() methods [11].

6.3   OpenCV and iOS

In order to incorporate the rich functionality provided by OpenCV into your iOS app, you need to learn how to: (1) integrate OpenCV into your project; (2) handle conversions between image objects in iOS (UIImage or CIImage) and OpenCV (cv::Mat); and (3) create a wrapper around the OpenCV functionality exposed to your app's code.

The three most common options to integrate OpenCV into an iOS project are [11]:

1. Download the official OpenCV framework for iOS ("iOS pack") from the official OpenCV website [7] and add the framework to your project.
2. Get the source files from GitHub [8] and build the library on your own according to the instructions in [3].
3. Use CocoaPods dependency manager [10].

In the remainder of this section we will follow option 1 (above). The block diagram in Fig. 6.1 shows how your app's Swift code can be integrated with the Objective-C/C++ code of OpenCV through a bridging header. The OpenCV framework will be included in the app, inside a wrapper written by you [16].

The main steps to integrate OpenCV into an iOS project are:[1]

1. Add the OpenCV framework to your project
2. Create the wrapper
3. Create a bridging header
4. Create a prefix header
5. Implement the OpenCV functionality you need
6. Use those methods from Swift

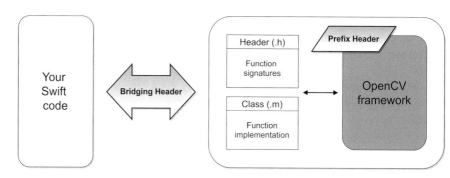

Fig. 6.1 Integrating OpenCV's functionality into an iOS app (redrawn from [16])

---

[1]See [16] for implementation details.

```
// From: https://github.com/jon-w1/OpenCV_demo
+ (cv::Mat)cvMatFromUIImage:(UIImage*)image{
  CGColorSpaceRef colorSpace =
  CGImageGetColorSpace(image.CGImage);
  CGFloat cols = image.size.width;
  CGFloat rows = image.size.height;

  cv::Mat cvMat(rows, cols, CV_8UC4);
  // 8 bits per component, 4 channels (color channels + alpha)
  CGContextRef contextRef = CGBitmapContextCreate
  (
  cvMat.data,       // Pointer to data
  cols,             // Width of bitmap
  rows,             // Height of bitmap
  8,                // Bits per component
  cvMat.step[0],    // Bytes per row
  colorSpace,       // Color space
  kCGImageAlphaNoneSkipLast | kCGBitmapByteOrderDefault
  // Bitmap info flags
  );

  CGContextDrawImage(contextRef, CGRectMake(0, 0, cols, rows),
    image.CGImage);
  CGContextRelease(contextRef);
  return cvMat;
}
```

Listing 6.1 Utility method to convert an UIImage to a cv::Mat (from [6])

One of the key differences when working with OpenCV in the iOS SDK is that you must first convert the UIImage object to an cv::Mat structure for use with an OpenCV library function and subsequently convert it back to a UIImage before attempting to display it on the screen.

Listing 6.1 shows sample code to convert from UIImage to cv::Mat [6] whereas Listing 6.2 shows sample code to convert back from an cv::Mat to a UIImage [6].

In order to access OpenCV's C/C++ code in your Objective-C/Swift environment, you must create a wrapper file that abstracts the OpenCV library behind relevant Objective-C method calls. Such wrapper consists of two files [6]:

- a header file called OpenCVWrapper.h, in which you should add interface method(s) for what your app will need to use OpenCV for.
- an implementation file called OpenCVWrapper.mm, in which you should be able to access the C++ header file from OpenCV and add any OpenCV-powered functionality required by your app.

Listings 6.3 and 6.4 shows sample Objective-C wrapper code [6] for a simple operation (RGB to grayscale conversion) using OpenCV's cv::cvtColor method.

```
// From: https://github.com/jon-w1/OpenCV_demo
+ (UIImage *)UIImageFromCVMat:(cv::Mat)cvMat{
  NSData *data = [NSData dataWithBytes:cvMat.data
    length:cvMat.elemSize()*cvMat.total()];

  CGColorSpaceRef colorspace;

  if (cvMat.elemSize() == 1) {
    colorspace = CGColorSpaceCreateDeviceGray();
  }
  else {
    colorspace = CGColorSpaceCreateDeviceRGB();
  }

  CGDataProviderRef provider =
    CGDataProviderCreateWithCFData((__bridge CFDataRef)data);

  // Create CGImage from cv::Mat
  CGImageRef imageRef = CGImageCreate(cvMat.cols, cvMat.rows, 8,
    8 * cvMat.elemSize(), cvMat.step[0], colorspace,
    kCGImageAlphaNone | kCGBitmapByteOrderDefault, provider,
    NULL, false, kCGRenderingIntentDefault);

  // get UIImage from CGImage
  UIImage *finalImage = [UIImage imageWithCGImage:imageRef];
  CGImageRelease(imageRef);
  CGDataProviderRelease(provider);
  CGColorSpaceRelease(colorspace);
  return finalImage;
}
```

Listing 6.2   Utility method to convert a cv::Mat to an UIImage (from [6])

```
// OpenCVWrapper.h
// From: https://github.com/jon-w1/OpenCV_demo

#import <Foundation/Foundation.h>
#import <UIKit/UIKit.h>

@interface OpenCVWrapper : NSObject

+ (UIImage *)grayscaleFromImage:(UIImage*)image;

@end
```

Listing 6.3   Sample OpenCV wrapper (.h file) (from [6])

```
//   OpenCVWrapper.mm
//   From: https://github.com/jon-w1/OpenCV_demo

#import <OpenCV/opencv2/opencv.hpp>
#import "OpenCVWrapper.h"

@implementation OpenCVWrapper

/*
 "Public" method available to outside callers
 */
+ (UIImage *)grayscaleFromImage:(UIImage*)image{
    cv::Mat matrix = [self cvMatFromUIImage:image];

    cv::Mat resultMatrix;

    /*
     * Add OpenCV method calls for processing/filtering
     */

    cv::cvtColor(matrix, resultMatrix, CV_BGR2GRAY, 4);

    // convert modified matrix back to UIImage
    return [self UIImageFromCVMat:resultMatrix];
}
```

Listing 6.4 Sample OpenCV wrapper (.mm file) (from [6])

Additional Recommendations

Once you get the OpenCV functionality integrated with your iOS app, you might still have to decide on which image processing operations to implement in Swift (as opposed to using their OpenCV functionality). According to [16], some OpenCV's functions (e.g., converting an image to grayscale) are typically much faster than the equivalent Swift implementations, but there are exceptions (e.g., blurring an image is many times faster in Swift than in OpenCV).

You should avoid passing images back-and-forth between Swift and OpenCV, due to the overhead involved in changing image formats (to/from OpenCV cv::Mat from/to UIImage or CIImage) which may slow down your app and lead to memory leaks [16].

As the functionality of Apple's Core Image and Vision frameworks continues to increase, the need for integrating OpenCV into iOS apps is likely to decrease, with the exception of highly specialized functionality (only available in OpenCV) or much faster implementations of certain functions, which would be worth the overhead and extra steps described in this section.

6.4   Learn More About It

The official OpenCV website includes rich documentation [5], tutorials [9], and examples for OpenCV-iOS integration [4].

The integration of OpenCV and iOS has received book-length treatment. See [13] and [15], for example.

The blog posts by Berrios and Williamson [12] and Poulsen [16] provide step-by-step instructions and complete examples of iOS and OpenCV integration.

References

1. OpenCV. https://opencv.org/about/. Accessed: 2020-07-10.
2. OpenCV 4.3.0. https://opencv.org/opencv-4-3-0/. Accessed: 2020-07-10.
3. OpenCV installation in iOS. https://bit.ly/3elCUhj. Accessed: 2020-07-10.
4. OpenCV iOS. https://bit.ly/3fi4Wvr. Accessed: 2020-07-10.
5. OpenCV modules. https://bit.ly/2Zj92ya. Accessed: 2020-07-10.
6. OpenCV on iOS: Demo. https://bit.ly/2WqaUDt. Accessed on 2020-07-11.
7. OpenCV: Releases. https://www.opencv.org/releases.html. Accessed: 2020-07-11.
8. OpenCV Source—GitHub. https://github.com/opencv/opencv. Accessed: 2020-07-10.
9. OpenCV Tutorials. https://bit.ly/3gRW6oV. Accessed: 2020-07-10.
10. CocoaPods. https://cocoapods.org/, 2018. Accessed: 2020-07-11.
11. E. Kurutepe. Face Recognition with OpenCV. https://bit.ly/3iVJwGG, 2015. Accessed: 2020-07-11.
12. F. Berrios and J. Williamson. Image Recognition in Apps: Embedding OpenCV in Android & iOS. https://bit.ly/30lgfly. Accessed: 2020-07-11.
13. J. Howse. *iOS application development with OpenCV 3: create four mobile apps and explore the world through photography and computer vision.* Packt Publishing, Birmingham, UK, 2016.
14. A. Kaehler and G. Bradski. *Learning OpenCV 3: Computer Vision in C++ with the OpenCV Library.* O'Reilly Media, Inc., 2017.
15. K. Kornyakov and A. Shishkov. *Instant OpenCV for IOS: Learn how to Build Real-time Computer Vision Applications for the IOS Platform Using the OpenCV Library.* Instant—short, fast, focused. Packt Publishing, 2013.
16. T. Poulsen. Using OpenCV in an iOS app. https://bit.ly/3fgBgPq. Accessed: 2020-07-10.

Printed in the United States
By Bookmasters